Sucker Nation

Vincent F. Hendricks

Sucker Nation

The Philosophy of Trading Wits for Vanity

Springer

Vincent F. Hendricks
Center for Information
and Bubble Studies
University of Copenhagen
København S, Denmark

ISBN 978-3-032-22787-4 ISBN 978-3-032-22788-1 (eBook)
https://doi.org/10.1007/978-3-032-22788-1

This Springer imprint is published by the registered company Springer Nature Switzerland AG
The registered company address is: Gewerbestrasse 11, 6330 Cham, Switzerland

Vincent F. Hendricks

Photo | Jan Grarup | ©2022

Cover design by Nanna Grunwald

Portrait photo by Jan Grarup

ABOUT THE AUTHOR

Vincent F. Hendricks is Professor of Formal Philosophy at The University of Copenhagen. He was the founder and director of the Center for Information and Bubble Studies (CIBS) funded by the Carlsberg Foundation. A prolific writer he has been awarded several prizes for his research among them The Elite Research Prize by the Danish Ministry of Science, Technology and Innovation, The Roskilde Festival Elite Research Prize, Choice Magazine Outstanding Academic Title Award and The Rosenkjær Prize. He was Editor-in-Chief of *Synthese: An International Journal for Epistemology, Methodology and Philosophy of Science* between 2005-2015.

Hendricks has over the years appeared in a wide variety of media outlets including The Washington Post, The Huffington Post, The Guardian, BBC, Le Monde, DIE ZEIT, Welt, TEDx, The Open Mind: Channel Thirteen, Wired, The Conversation, VICE, Quartz, New Statesman, Frankfurter Allgemeine, Danish Broadcasting Company, News POLSAT Poland, The Tom Hartman Show, The

Pete Dominick Show, The David Packman Show, The Forbidden Speech podcast, SiriusXM, Ideasphere – National Public Radio, Sputnik-News Radio UK, SRF Tagesschau, Neue Zuricher Zeitung, CNBC AFRICA, New Business Ethipoia, CPH:DOX, TRT World Turkey, ARD Radio Germany, TV2 News Denmark.

Contents

For the clan

—Lady Day, Master Isak, Princess, Milster &
Lille-Lille—

PROLOGUE

"Only don't tell me you're innocent. Because it insults my intelligence and makes me very angry," is a famous quote from Michael Corleone in the legendary movie *The Godfather*. Michael relays the message to his brother-in-law, Carlo Rizzi, suspected of being involved in the murder of Michael's brother, Santino. Michael's statement implies his belief that Carlo's lie of innocence is so obvious that it would be insulting, and so he is extraordinarily angry about Carlo's attempt to deceive Michael. Carlo is the hustler taking Michael for the sucker.

"Don't be a sucker" has become one of the quiet commandments of modern life. It is spoken half as advice, half as accusation. To be a sucker is not merely to lose, but to lose because one failed to see how things really work. The warning implies that the world is a game, that someone is always hustling, and that survival depends on learning the rules before they are used against you. *Everything is a hustle.*

Yet this warning conceals a deeper contradiction. The very qualities that make social life possible—trust, sincerity, openness, participation—are the very same features making suckerhood possible. Without them, nothing holds together. With them, everything becomes exploitable. Modern society does not resolve this contradiction but merely organizes itself around it.

The distinction between hustlers and suckers is therefore not about intelligence or morality. It is structural. Hustlers and suckers are not different kinds of people but different positions in a social world where visibility is required in order to act, attention is scarce, and interpretation is thin. To appear at all—to speak, to matter, to participate, even to be—one must expose oneself to systems that do not recognize persons, only signals.

The sucker is the agent who continues to act *as if* appearance were still a form of recognition, *as if* sincerity would be met as sincerity, *as if* freedom of speech is really free, *as if* waiting in line at high-end coffee joints, hyped restaurants or fancy clubs mirrors a coffee connoisseur, a foodie or a celebrity—*as if* participation itself had intrinsic meaning. The hustler is the agent who knows that appearance and meaning have come apart, and who lives strategically in that knowledge. One trades wits for vanity; the other preserves wits by exploiting vanity. Both positions carry costs. Neither offers innocence.

What makes this condition uniquely modern is not exploitation itself, but its scale and impersonality. Platforms, markets and media systems no longer need to deceive. They only need people to keep *showing up* as Sam "Ace" Rothstein conveys in another epic movie, *Casino*: "In the casino, the cardinal rule is to keep them playing and to keep them coming back. The longer they play, the more they lose, and in the end, we get it all." Attention, outrage, authenticity, personal information, even critique—all may be extracted and converted into value. Refusal, by contrast, risks invisibility—and when invisible, you are a sure nobody in attention economy.

This is why insight alone does not dissolve the problem. Many people know they are being exploited, yet continue to participate. They do not know whether others are ready to stop paying the cost of being seen. Without that knowledge, withdrawal feels futile, and silence feels like disappearance. Agency itself becomes conditional on exposure in the ever growing attention economy ruling the information market at large.

This book does not promise a way out of this dilemma. There is no clean exit from a world in which appearing is both necessary and dangerous. What it offers instead are the nuts and bolts for understanding the trade we are already making—often without naming it—between strategic intelligence and recognition, between distance and belonging, between not being a sucker and still having a face in the world. *Sucker Nation* offers a systematic framework for understanding why we often enough decide to trade wits for vanity.

Vincent F. Hendricks | Copenhagen | March | 2026

CHAPTER 1: Hustlers, Suckers and the Price of Being Seen

Modern social life is increasingly organized around a distinction that is at once familiar and yet poorly understood: The distinction between *hustlers* and *suckers*. To hustle is to be savvy, strategic, alert to hidden incentives yet to be a sucker is to be exposed, taken in, left paying for someone else's gain. The terms circulate easily in contemporary discourse—in finance, politics, digital culture and around the dinner table—but their apparent clarity conceals a deeper structure. Hustlers and suckers are not merely types of people. They are positions generated by the epistemic and moral economy of modern society.

1.1 Sucker Nation

The simple but unsettling claim of *Sucker Nation* is that modern agency requires *trading wits for vanity*. To re-

V. F. Hendricks, *Sucker Nation*,
https://doi.org/10.1007/978-3-032-22788-1_1

main socially present, morally legible, and politically relevant, individuals must expose themselves—display sincerity, identity, outrage, or authenticity—under conditions where such exposure is systematically converted into extractable value. Those who accept this trade become suckers, not because they are foolish, but because they have surrendered strategic distance in exchange for recognition. Those who refuse it, or exploit it, become hustlers—not necessarily because they are more intelligent or apt—but because they have figured out how to preserve their wits while living off the vanity of others.

Vanity here is not narcissism or some sense of excessive self-love (Lasch, 1979). It is the belief that being seen as one truly still matters—that sincerity will be recognized as sincerity, that expression will be interpreted as intended, that participation will count as agency rather than merely as input. This belief is not immediately just irrational. It may very well be inherited from social worlds of the past in which audiences were limited, infrastructures more managable, interpretation was richer and trust had a chance of being reciprocated.

1.2 ... *as if* ...

In modern platform societies these beliefs of old may become a downright liability. Visibility is no longer primarily a mode of recognition but a resource to be extracted (Srnicek 2016) Expression is not interpreted as much as simply aggregated (McNamee 2020). Meaning is not evaluated but rather processed—and the faster the better with little to no friction only to stimulate engagement, traffic and data harvest (Couldry, N. & Mejias, U.A. 2019, Hendricks & Mehlsen 2022). Under these conditions, to act openly and sincerely is to give up strategic intelligence—*wits*—in return for the promise of acknowledgment or recognition, in short—*vanity*. The trade is rarely explicit, but it is everywhere enforced and everywhere to be found.

The sucker is the agent who accepts this exchange. They act *as if* sincerity protects meaning, *as if* moral expression resists manipulation, *as if* there is freedom of speech is without limits and *as if* participation itself alters collective belief. Now, the hustler understands that this is not so. Hustlers know that visibility and recognition are decoupled, that outrage travels regardless of truth, understands that whatever goes viral is not necessarily true, and what is true does not necessarily go viral. Hustlers have caught on to the idea that authenticity is not an antidote to exploitation but actually one of its most efficient inputs.

Surprisingly, the distinction between hustlers and suckers is thus *epistemic* long before it is moral. What separates them is not intelligence or virtue, but *distance*—distance between belief and expression, identity and visibility, participation and commitment—and the cognitive dissonance these distinctions invite. The sucker chooses to collapse these distinctions to overcome the dissonance, believing that to speak sincerely is to act meaningfully. The hustler preserves them, treating sincerity as a style, visibility as a tool, and participation as something to be either optimized or consciously avoided if need be.

1.3 No Common Knowledge

Never in the history of mankind have there been so much information readily available—information in abundance is the name of the game these days. The transition to an age of abundance has significant effects on how humans process information. When the supply exceeds human processing capacity, it may lead to "infoxication" or "information obesity" causing everything from stress, anxiety to difficulties in making decisions (Shahrzadi et al. 2024). The paradox of abundance is that increased access to information may sometimes make people *less* well-informed if they cannot effectively synthesize the data or distinguish between high-quality facts and information pollution (Lynch 2016, Gogoi et al. 2024). Thus the vast, readily accessible supply of content available through various sources and devices does not necessarily imply that the basis for *individual* deliberation, decision and action has improved accordingly. Nor does it imply that *collective* deliberation, decision, action and coordination has improved just because information is in abundance.

Much of social life depends not merely on what individuals know, but on *what they know about what others know*. This higher-order dimension of knowledge is captured by the concept of *common knowledge* (Lewis, 1969). A fact is common knowledge in a group when everyone knows it, everyone knows that everyone knows it, everyone knows that everyone knows that everyone knows it, and so on forever after. While this infinite regress may sound abstract, it describes something deeply practical. The shared background that allows coordination, trust, and cooperation to function without constant verification (Pinker, 2025). Information in abundance, does not entail common knowledge in abundance either.

That is indeed unfortunate, because common knowledge is what allows us to cross an intersection without negotiation, follow conventions without contracts, and cooperate without continually testing one another's intentions. When a rule, norm, or expectation is common knowledge, individuals can rely on it confidently, knowing that others will do the same (Lewis, 1969). Coordination becomes cheap, trust becomes reasonable, and cooperation becomes stable: "Our world is built on conventions that allow us to coordinate effectively and are self-reinforcing because they are common knowledge." (Pinker 2025: 201) When common knowledge breaks down, these capacities erode—even if individuals remain intelligent, informed, and well-intentioned.

It is of paramount importance to distinguish individual knowledge from common knowledge. Individuals know a great deal:

"We know a lot"

are the very first words of David Lewis' seminal paper "Elusive Knowledge" (Lewis, 1996). We may know that markets are speculative, that social platforms manipulate attention, that our data is the new oil, that all data collected is credit information, that outrage is profitable, or that sincerity is often exploited—know that we have surrendered hook, line and sinker to the attention economy. Yet knowing these things individually does not suffice for concert deliberation, decision and action. What matters is whether individuals know that others know them, and know that others know that they know.

Without common knowledge, agents must act under uncertainty about others' expectations. This uncertainty may infect even the most rational of decisions. One may suspect that cooperation is fragile, but still cooperate because one does not know whether others are ready to stop cooperating. One may recognize manipulation of attention, but continue to participate because one does not know whether withdrawal would be shared or isolated.

This is exactly the epistemic environment in which the hustler/sucker distinction emerges. The distinction is not primarily about misinformation or gullibility. It is rather about *higher-order ignorance*—about who knows what, and who knows that others know it. The sucker is not someone who lacks knowledge. Rather, the sucker is once more someone who acts *as if* trust, sincerity, free speech or cooperation are still common knowledge, even when they are not. The hustler is someone who recognizes that these assumptions no longer hold, and who exploits the resulting asymmetry.

This exploitation does not require deception. It requires only that some agents continue to behave *as if* certain norms are shared, while others know that they are not. Because this asymmetry is not itself common knowledge, it persists. The sucker cannot safely withdraw without risking isolation, and the hustler cannot be easily identified or collectively sanctioned. In this way, the absence of common knowledge stabilizes exploitation. Trust depends not only on beliefs about others' character, but on beliefs about others' expectations. When common knowledge erodes, trust becomes fragile. Individuals may still *want* to trust, but they cannot be confident that trust will be reciprocated. As a result, trust becomes selective, strategic or feigned and performative.

Ironically, this does not lead immediately to universal distrust. Instead, it leads to a *divided* social world in which some agents continue to trust as a condition of participation, while others strategically refrain. The former are vulnerable to exploitation while the latter are insulated but alienated. The social fabric frays unevenly.

It has been long established that coordination problems become especially acute in the absence of common knowledge (Schelling 1990, Ostrom 1990, Thomas et al. 1991, De Freitas et al. 2019, Pinker 2025). Even when all parties would benefit from changing behavior—say, by refusing to participate in exploitative social platforms—coordination often fails. Each agent waits for others to move first, uncertain whether withdrawal will be collective or solitary. The result is inertia. This explains why practices widely recognized as harmful can persist for long periods. Individuals know the costs, but cannot coordinate around alternatives because the epistemic conditions for collective action are missing. Without common knowledge, even shared interests fail to produce shared action.

Modern, mediated societies are particularly prone to common knowledge failure. Scale, anonymity, algorithmic mediation, and fragmented publics make it difficult to know what others believe or expect. Signals of consensus—likes, stories, trends, prices, outrage—often reveal attention rather than belief (Hendricks & Mehlsen 2022, Berger, Moe & Schweidel 2023). They obscure rather than clarify higher-order knowledge. As a result, social life increasingly takes place in an *epistemic fog.* People know a lot, but do not know what can be relied upon as shared. This fog sustains the hustler/sucker distinction while undermining the very capacities—trust, coordination, cooperation—that social life requires. This asymmetric curse is *no–common knowledge.*

The hustler/sucker distinction is a symptom of life in a world where information proliferates but common knowledge withers—and where acting together becomes harder precisely as individuals become more informed. Thus understanding the distinction between hustler and sucker, between common knowledge and lack thereof is not merely an intellectual exercise. It is a way of diagnosing why modern societies struggle to coordinate around shared problems, why trust feels risky, and why cooperation so often collapses under the weight of knowing too much, but not knowing together (Hendricks & Hansen, 2016).

1.4 The Tragedy of Modern Agency

In contemporary attention culture, exposure is praised as courage, authenticity as virtue, and vulnerability as strength. Strategic restraint, by contrast, is often stigmatized as cynicism or bad faith. The sucker is therefore not merely tolerated but celebrated, while the hustler disavows their own position even as they benefit from it. That's moral inversion.

This inversion deepens the tragedy. Individuals do not only trade wits for vanity—they trade them for *moral innocence* as well. To refuse exposure feels not only risky but wrong. To remain silent feels like complicity. The cost of strategic intelligence becomes shame, while the cost of exposure is normalized as a downright ethical commitment.

The hustler/sucker distinction reveals an indeed very tragic structure at the heart of modern agency. To act at all, one must trust that expression matters. To avoid exploitation, one must simultaneously distrust the very conditions that make expression possible. There is no position that preserves both full agency and full security. Insight does not dissolve this conflict but rather sharpens it.

The tragedy is not that some people are hustled while others are not. The tragedy is that *agency itself now requires a willingness to be hustled*, and that those who refuse this condition must accept marginalization, opacity or isolation as the price of keeping their wits. The price of individual autonomy and self-determination honestly becomes so high, that no sucker is willing to pay it. Then rather a sucker sell-out of wits for vanity. That's precisely what the hustler exploits. That's the hustler's game for suckers.

CHAPTER 2: The Hustler's Game for Suckers

The distinction between hustlers and suckers is usually treated as informal, even vulgar—a matter of street wisdom rather than theory. Hustlers are clever, suckers naïve; one exploits, the other is exploited. Such descriptions invite moral judgment or psychological diagnosis, while obscuring something inherently more troubling: The fact that hustlers and suckers may arise *even when all agents are intelligent, reflective, and instrumentally rational.*

2.1 Turn to Game Theory

This initial observation motivates the turn to game theory. The aim of game theoretical formalization is not to sanitize a colloquial distinction, but to show that the hustler/sucker divide is not an accident of character. It is a *structural possibility of strategic interaction under epis-*

V. F. Hendricks, *Sucker Nation*,
https://doi.org/10.1007/978-3-032-22788-1_2

temic asymmetry. Consulting game theory, especially in its epistemic and informational variants (Aumann 1976, 1999), (Brandenburger 1992), (Aumann & Brandenburger 1995), (Galeazzi, Pacuit & Roy 2026) is a good bet to capture this very structure.

At a first approximation, the hustler/sucker distinction appears paradoxical. In standard game-theoretic models, rational agents *best-respond* to incentives (Hyland et al, 2024). If an action is exploitable, rational players should avoid it. If exploitation indeed persists, it seems to indicate irrationality or error. And yet in many real-world interactions like markets, bargaining, signaling environments, (social) platform economies—exploitation is stable, predictable, and often asymmetric in a very persistent way. The same agents repeatedly occupy the hustler role, while others repeatedly absorb losses, despite learning, experience, and immediate awareness. How may exploitation be compatible with rationality? And why does exploitation not automatically disappear through learning, experience or selecting some some proper equilibrium?

Game theory invites an immediate conceptual clarification of significance. "Hustler" and "sucker" cannot be defined as merely types of persons with certain characteristic traits. They must be defined as *positions within a game*—as roles determined by strategies, beliefs, and information structures in which the agents operate. In such a framework, suckers are not just whoever fail to maximize expected utility given their beliefs. The suckers may indeed be *subjectively rational* as they best-respond to the beliefs they hold. Their vulnerability arises because those beliefs are systematically mistaken about the strategic environment—typically about others' incentives, rationality, or future actions for that matter.

Conversely, a hustler is not someone who violates rationality or some predefined equilibrium constraints. The hustler is *fully rational.* What distinguishes hustlers is that their optimal strategy depends on others having false or partial beliefs, and on those belief asymmetries not being common knowledge. Game theory thus reframes the distinction in such a way that the sucker plays a strategy which is optimal under false or incomplete beliefs while the hustler plays a strategy which is optimal *because* others hold such faulty beliefs.

If all players knew that others were rational and strategically sophisticated—and knew that this was known, and so on *ad infinitum*—then many sucker strategies would be eliminated by standard solution concepts of game theory (like rationalizability or sequential equilibrium). This is in essence the insight behind results linking common knowledge of rationality and correct beliefs or epectations to equilibrium play in game theory.

The hustler/sucker distinction therefore lives precisely in the gap left by these results. That is, situations in which rationality is *not* common knowledge, or in which agents assign positive probability to naïve, cooperative, or norm-following types. In such settings, a player can rationally choose a strategy that exploits the expected mistakes of others, while those very others may rationally choose strategies that are exploitable given their beliefs. Game-theoretically, this moves focus away from payoff matrices alone and toward type spaces, belief hierarchies, off-equilibrium beliefs, and the limits of equilibrium refinements. The distinction between hustler and sucker is thus not an anomaly but a *prediction of epistemic game theory* once common knowledge assumptions are relaxed.

2.2 "Somebody's a sucker"

Hustler/sucker interactions are not necessarily unstable or short-lived. In many models, such unfortunate interactions can persist in equilibrium—not because agents fail to learn, but because learning itself is constrained by the (epistemic) structure of the game. If belief asymmetries are reinforced by the game's own outcomes—if sucker behavior appears to "work" often enough, or if hustler strategies conceal their exploitative nature—then Bayesian updating does not necessarily eliminate exploitation. Instead, it may actually stabilize it.

Here is a disturbing game theoretical implication: *There may be no equilibrium in which everyone avoids being a sucker.* In everyday language, an equilibrium suggests balance or stability. In game theory, the idea is more precise but also more subtle. An equilibrium is not a morally desirable outcome, nor a fair one, nor even a harmonious one. It is a situation in which *no agent has a unilateral reason to change their behavior*, given what they believe about others and given what others are doing (Nash 1951,

Harsanyi & Selten 1988, Schelling 1990, Myerson 1991). This definition is key for understanding the hustler/sucker distinction. Many people assume that if an arrangement is exploitative, it cannot be stable—surely the exploited party would stop participating once they realize what is happening. Game theory shows why this intuition is often wrong.

At its core, an equilibrium is a configuration of strategies such that *each player's strategy is a best response to the strategies they expect others to play.* Importantly, "best response" is always relative to beliefs. An agent may be fully rational and still play an exploitable strategy if, given their beliefs, no alternative strategy looks better. In the hustler/sucker setting:

- The *sucker* plays a strategy (e.g., trusting, cooperating, revealing, participating) that is optimal given their beliefs about others' behavior.

- The *hustler* plays a strategy (e.g., exploiting, free-riding, rage baiting, extracting attention, timing exit) that is optimal given their belief that others will continue to play exploitable strategies.

If neither party sees a unilateral deviation that would improve their outcome—given what they believe—then the interaction is in equilibrium, even if it is asymmetric and unfair.

The key mistake in "don't be a sucker"-everyday thinking is to assume that exploitation requires irrationality. In fact, suckerhood may be perfectly rational. Suckers believe that most others are trustworthy, cooperative, or sincere, may believe that withdrawing trust would impose higher costs (exclusion, invisibility, lost opportunity) than continuing to participate and they may even believe that exploitation is rare, accidental, or temporary. Given these beliefs, continuing to play the "sucker" strategy can be optimal. These beliefs may very well be supported by limited experience, social norms, or incomplete feedback. As long as deviating—becoming cynical, withdrawing, refusing to participate—appears worse, the sucker has no reason to change. Their position is therefore *stable*, even if it is costly.

Consider the hustler's position. The hustler believes that some agents will continue to act trustingly or transparently. He or she believes that these agents will not immediately detect exploitation or coordinate withdrawal. And finally believes that exploiting these behaviors yields higher payoffs than cooperating. Given these beliefs, hustling is a best response. Crucially, the hustler's strategy depends on the expectation that sucker strategies will persist. If that expectation remains justified, the hustler has no incentive to stop. Thus, hustling and suckerhood *mutually reinforce one another*. Each role is rational given the other. An equilibrium in the hustler/sucker setting does not imply that players agree on the nature of the game, that players approve of the outcome nor that players are equally well off. It only implies *stability under unilateral deviation.*

The sucker may resent being exploited but see no better

alternative. The hustler may distrust the sucker but continue exploiting because it pays. Neither changes course because doing so alone would make them worse off. This is why equilibria can be deeply unjust, corrosive, or tragic—and still persist. In the hustler/sucker setting, equilibrium depends critically on higher-order beliefs about what agents believe about others' rationality, what they believe others believe and very importantly what they believe others believe about them.

As long as the sucker believes that withdrawal will not be widely shared, and the hustler believes that exploitation will not be anticipated, the equilibrium holds. This also explains why information alone often fails to disrupt exploitation. Knowing that exploitation exists is not enough. What matters is whether it becomes common knowledge that exploitation is anticipated and will be collectively resisted.

An equilibrium breaks when at least one player gains a reason to deviate unilaterally. In the hustler/sucker context, this can happen if suckers come to believe that others will also withdraw, or hustlers come to believe that exploitation will be detected and punished or belief asymmetries collapse through transparency, regulation, or shared awareness. At that point, sucker strategies are no longer optimal, hustler strategies lose their foundation, and the equilibrium dissolves.

Equilibrium in the hustler/sucker setting has a tragic dimension. The equilibrium persists not because it is good, but because *no one can escape it alone*. Individual rationality does not guarantee collective improvement. Stability is achieved at the cost of vulnerability, cynicism, or loss of trust. In this sense, an equilibrium names not a solution, but a bind: A situation in which rational agents remain locked into patterns they would collectively prefer to escape, but cannot. Understanding equilibrium in this way is essential for grasping why the hustler/sucker distinction is so resilient—and why it cannot be eliminated simply by becoming smarter, more informed, or more virtuous.

Under certain informational conditions, suckerhood is not a mistake but a role that must be occupied by someone if the game is to be played at all. This is the point at which the formal analysis connects back to the philosophical framing. What appears, informally, as stupidity or vanity may, formally, be the price of participation in a game whose rules require trust, optimism, or norm-following by at least some players.

Formalizing the hustler/sucker distinction does not strip it of its philosophical or social meaning. On the contrary, it sharpens it. By translating the distinction into the language of strategies, beliefs, and equilibrium concepts, game theory reveals that exploitation can be fully consistent with rational choice, insight alone does not guarantee escape and that structural vulnerability can persist even among sophisticated agents.

The formal models to follow should be read not as abstractions from social reality, but as *diagnostic instruments*. They show precisely how a world populated entirely by rational agents can still produce systematic asymmetries of advantage. The instruments also reveal why exhortations to "be smarter", "wise up" or "don't be a sucker" miss the point.

2.3 Formal Hustlers and Suckers

A hustler is a player who maximizes payoff by strategically exploiting systematic belief errors, informational asymmetries, or equilibrium-selection failures of other players, rather than by altering the underlying payoff structure. Formally, player i is a hustler in a game G with strategy σ_i and a belief profile μ_{-i} such that

$$\sigma_i \in \arg\max_{\sigma_i^{'}} \mathbb{E}_{\mu_{-i}} \left[u_i \left(\sigma_i^{'}, \sigma_{-i}\right)\right].$$

Unpacking the formula, arg max returns a set of strategies that achieve the highest value of the expression; $\sigma_i^{'}$ ranges over all strategies available to player i while

$$\mathbb{E}_{\mu_{-i}}[u_i(\sigma_i^{'}, \sigma_{-i})]$$

denotes the expected payoff to player i if player i were to play the (possibly hypothetical) strategy $\sigma_i^{'}$, while all other players play according to their mixed strategies μ_{-i}.

Suckers are *subjectively rational* in the sense that they choose a strategy that maximizes expected utility given their beliefs about the game and about others' behavior. Their vulnerability arises because these beliefs—often about trust, cooperation, or shared norms—are false or incomplete. A sucker is a player whose strategy is not payoff-maximizing σ_i given the true incentive structure of the game, but is optimal only under systematically false beliefs about other players' strategies, types, or rationality. Formally, player i is a sucker if there exists a game G and a belief μ_i such that

$$\sigma_i \in \arg\max_{\sigma_i^{'}} \mathbb{E}_{\mu_i}\left[u_i\left(\sigma_i^{'}, \sigma_{-i}\right)\right].$$

Thus player i chooses a strategy σ_i that is among the best possible choices, given what is believed about how the other players will act. The problem is that player i's strategy σ_i is *not* an optimal (best-response) strategy, given beliefs about what the other players will do, thus

$$\sigma_i \notin \arg\max_{\sigma_i^{'}} \mathbb{E}\left[u_i\left(\sigma_i^{'}, \sigma_{-i}\right)\right].$$

Given what player i expects others to do, σ_i is accordingly a mistake.

To characterize hustlers and suckers in epistemic game theoretic terms let:

i. T_i be the type space of player i,

ii. μ_i be beliefs over T_{-i},

iii. Rationality be common knowledge *only up to some finite epistemic depth.*

A hustler is a type $t_i \in T_i$ such that:

1. t_i is fully Bayesian-rational,

2. t_i assigns positive probability to irrational or naive types of others,

3. t_i chooses strategies optimal conditional on those belief distortions.

This breaks common knowledge of rationality while preserving individual rationality. To characterize the sucker, keep the framework above while substituting (iii)-condition for (iii)': Rationality fail to be common knowledge (also entailed by (iii)). Based on (i)-(iii)', a sucker is a type $t_i \in T_i$ such that:

1. t_i is subjectively rational (best-responds to beliefs),

2. t_i assigns excessive probability to cooperative, fair, or truthful types,

3. t_i fails to update beliefs correctly after informative deviations.

A hustler benefits from equilibria sustained by non-credible beliefs, misinterpreted signals and weak equilibrium which formally amounts to the fact that the hustler's maximizing payoff strategy σ_i lies in a *perfect Bayesian equilibrium* but not in a sequential equilibrium under correct belief revision.

Let $\mathcal{B}_{-i}$ be the set of beliefs held by others about player i and let BR_{-i} denote the best response of others. A hustler selects actions a_i for which it is possible that other players are acting rationally, given some beliefs they could reasonably have, i.e.

$$\exists\, \mu_{-i} \in \mathcal{B}_{-i} \text{ with } a_{-i} \in BR_{-i}(\mu_{-i})$$

even though others are making a mistake relative to player i's action:

$$a_{-i} \notin BR_{-i}(\sigma_i).$$

What the hustler does is to create informal mechanisms like promises, reputations or narratives. These mechanisms satisfy incentive compatibility for others *only* under false beliefs while the hustler remains incentive-compatible under *all* belief realizations. So again, a hustler is a rational player whose optimal strategy depends on others being systematically mistaken about the game they are playing.

In repeated games, the design remains the same in that a hustler mimics cooperative types in early periods, then exploits discounting or trust thresholds and finally defects optimally when belief convergence occurs with the switch timed to maximize discounted payoff given opponents' belief updating.

A sucker participates in equilibria sustained by overly optimistic beliefs, normative rather than strategic expectations and failure of equilibrium refinements. Formally the sucker's strategy σ_i is part of a perfect Bayesian equilibrium only under misspecified beliefs because under correct belief revision, the strategy is eliminated by sequential rationality.

Again, let $BR_i(\mu_i)$ denote best responses under belief μ_i. A sucker chooses action $a_i \in BR_i(\mu_i)$ such that

$$a_i \notin BR_i(\sigma_{-i}).$$

A sucker accepts informal mechanisms like promises, reputations, assurances and moral commitments which are not incentive-compatible under truthful play and participation is individually rational only under belief distortion.

In repeated interactions, a sucker maintains cooperation beyond the point of optimal defection, misinterprets finite histories as evidence of long-term commitment and fails to anticipate end-game exploitation. A sucker is a player whose optimality depends on the false assumption that others are playing a different game.

Given the game theoretical characterization, the hustlers and suckers are established as *epistemic duals* because

- Hustler: payoff from others' belief errors.
- Sucker: payoff loss from own belief errors.

Very well, this may be common knowledge now, but what do all these formalities have to do with the *lack* of common knowledge and the hustler/sucker distinction? That's next.

CHAPTER 3: No Common Knowledge in Sucker Nation

The hustler/sucker distinction appears, at first glance, to be an enduring feature of social life. In markets, politics, on social platforms and in everyday interactions, some agents seem consistently able to exploit others who appear persistently vulnerable. Game theory offers a precise diagnosis. It shows that the distinction between hustlers and suckers is not rooted in differences of intelligence or virtue, but in the absence of common knowledge of rationality.

3.1 Common Knowledge of Rationality

Rationality, in game theory, means that agents choose strategies that maximize expected utility given their beliefs. Importantly, rationality by itself does not imply that

V. F. Hendricks, *Sucker Nation*,
https://doi.org/10.1007/978-3-032-22788-1_3

agents know others are rational, or that others know that they know this, and so on. Common knowledge of rationality is a much stronger condition. It requires that all agents are rational, that all agents know that all agents are rational, that all agents know that all agents know that all agents are rational, and so forth *ad infinitum*. This infinite hierarchy is not an abstract flourish. It captures the condition under which agents can reliably reason about others' strategic reasoning without fear that their expectations rest on mistaken assumptions.

The hustler/sucker distinction depends on a very specific epistemic configuration. The sucker is an agent who plays a strategy that is optimal given their beliefs, but whose beliefs are false—typically about the trustworthiness, cooperation, or strategic restraint of others. The hustler is an agent whose strategy is optimal precisely because others hold such beliefs. Hustling is profitable only because some players are expected to behave in ways that a fully informed, fully alert agent would not.

This configuration requires that belief asymmetries are not common knowledge. The hustler must believe that others are mistaken, while those others must fail to know that they are mistaken, or fail to know that the hustler knows they are mistaken. The entire distinction rests on higher-order ignorance.

Under common knowledge of rationality, sucker strategies are eliminated by standard game-theoretical reasoning. Rational agents will not assign positive probability to others playing strategies that are strictly dominated or exploitable once rationality is assumed at every level. In such a setting, a player cannot rationally expect others to behave naively, cooperatively, or trustingly if doing so would expose them to exploitation. Since everyone knows that everyone knows this, no one can rationally plan to exploit such behavior either. Sucker strategies become irrational not because they are morally flawed, but because they rely on beliefs that are inconsistent with common knowledge of rationality. If sucker strategies disappear, so too does hustling. Hustler strategies are best responses to the expectation that others will act in exploitable ways. Once such expectations are no longer rationally permissible, hustling loses its strategic foundation and advantage.

Under common knowledge of rationality, attempting to hustle would itself signal irrational expectations. An agent who plans to exploit others must believe that others will fail to anticipate exploitation. But this belief contradicts common knowledge of rationality. As a result, hustler strategies are eliminated alongside sucker strategies. There is no one left to exploit, and no rational basis for expecting exploitability. In this sense, hustlers and suckers disappear together. They are not opposing types that cancel one another out. They are mutually sustaining roles that collapse once the epistemic conditions that support them are removed. That's the basic informal idea of how the hustler/sucker distinction disappears under common knowledge.

3.2 Proof in Sketch

To be more formal about the disappearance act of the hustler/sucker distinction under common knowledge of rationality, let G be a finite normal-form game with players $i = 1, \ldots, n$, action sets A_i, utilities u_i. An epistemic model specifies:

- a state space Ω,
- for each player i: An information partition Π_i,
- and beliefs $P_i(\cdot \mid \Pi_i(\omega))$ at each state ω.

Let $a_i(\omega) \in A_i$ be the action actually chosen by player i in state ω. Next, define the event that player i is rational, R_i, at ω:

$$R_i := \{\omega \in \Omega : a_i(\omega) \in \arg\max_{a_i \in A_i} \mathbb{E}_{P_i(\cdot|\Pi_i(\omega))}[u_i(a_i, a_{-i}(\cdot))]\}.$$

Ω is the state space encoding the actions chosen by all players, their information and their beliefs. $\Pi_i(\omega)$ denotes everything that player i knows at ω while $P_i(\cdot|\Pi_i(\omega))$ is player i's belief (probability distribution) over states, conditional on their information. Finally $a_{-i}(\cdot)$ is the action profile of the other players as a function of the state and

$$\mathbb{E}_{P_i(\cdot|\Pi_i(\omega))}[u_i(a_i, a_{-i}(\cdot))]$$

is player i's expected payoff from choosing action a_i, given the beliefs about what the other players will do, based on their information at ω.

Let

$$R := \bigcap_i R_i$$

be "everyone is rational" —thus all players are acting rationally. To define common knowledge of rationality ($CK(R)$) let $K_i(A)$ denote the event "player i knows A" or the set of states where player i knows that proposition A is true. Common knowledge of R is the event

$$CK(R) \ := \ \bigcap_{m=1}^{\infty} K^m(R),$$

where

$$K^1(R) := \bigcap_i K_i(R)$$

and

$$K^{m+1}(R) := \bigcap_i K_i(K^m(R)).$$

To unpack, $K^1(R)$ refers to the first level where everyone knows that everyone is rational. That is player 1 knows everyone is rational, player 2 knows everyone is rational, and so on for all players. It is mutual knowledge, but not yet common knowledge. $K^{m+1}(R)$ refer to higher levels where everyone knows that everyone knows that … (m times) … everyone is rational. Each iteration adds one more layer of "knowing that others know". This sequence builds higher-order epistemic certainty where R is everyone is rational; $K^1(R)$ is everyone knows everyone is rational; $K^2(R)$ everyone knows that everyone knows everyone is rational ... all the way up to

$$\bigcap_{m=1}^{\infty} K^m(R)$$

(full) common knowledge of rationality (in the limit ∞). With these preliminaries in place it is now possible to prove the following theorem in three steps:

Theorem: *Hustlers, Suckers and Common Knowledge of Rationality*

> *On any state $\omega \in CK(R)$, neither "suckers" nor "hustlers" can exist.*

Proof: **The first step** is to show that common knowledge of rationality implies that actions are *rationalizable* (Bernheim 1984, Pearce 1984). This roughly means, that once everyone is known to be rational, no one will play a strategy that can be ruled out as unreasonable given that assumption. A standard result in epistemic game theory is that common knowledge of rationality implies rationalizability—or more precisely—that common knowledge of rationality is logically equivalent to the set of rationalizable strategies (Chin-Chiu Tan, T. & da Costa Werlang, S. R. 1988).

In the current setup, if $\omega \in CK(R)$, then for every player i, the action $a_i(\omega)$ is *rationalizable* (i.e., survives iterated elimination of never–best responses / is a best response to some belief over opponents' rationalizable actions): If everyone is rational and this is common knowledge, then each player believes others choose best responses to *some* beliefs. But since everyone knows that everyone knows ... common knowledge of rationality, the support of beliefs must lie in actions that are themselves consistent with rationality at all epistemic depths. Iterating this reasoning yields exactly the fixed point called *rationalizability*. Thus, common knowledge of rationality shrinks the set of possible play to rationalizable actions only.

Sucker actions are often rational enough, but only under beliefs that others will behave naïvely, cooperatively, or non-strategically. Under common knowledge of rationality, such beliefs are no longer admissible. Since everyone knows that everyone is rational, no one can rationally expect others to behave in systematically exploitable ways. As a result:

- *sucker actions cease to be rationalizable*, because they rely on beliefs that contradict common knowledge of rationality,

- *hustler actions also cease to be rationalizable*, because they rely on the expectation that sucker actions will occur.

Both roles disappear, not because agents become morally better or cognitively smarter, but because the epistemic assumptions that made those actions defensible are no longer coherent.

The second step is to show that suckers disappear under common knowledge of rationality. Recall that a sucker is a player who plays an action that is optimal only under *systematically false beliefs* about others' play/types, and in particular *fails to best respond* to the actual strategic structure when beliefs are correct. To make this precise in the epistemic model, define a sucker at ω as one who plays an action that is *not rationalizable*:

$$Sucker_i(\omega) \ := \ a_i(\omega) \notin \mathrm{Rat}_i(G),$$

where $\mathrm{Rat}_i(G)$ is the set of rationalizable actions for player i. From here, the proof is immediate: If state $\omega \in CK(R)$, then by common knowledge of rationality implying rationalizability (the first step), $a_i(\omega) \in \mathrm{Rat}_i(G)$ for all i. Therefore

$$Sucker_i(\omega)$$

is *false* for all i. Under common knowledge of rationality, nobody can be a sucker (in the belief-error/exploitability sense), because any realized action must be rationalizable.

The third step is to prove that also hustlers disappear under common knowledge of rationality. Now, a hustler is a player who profits from others' belief errors / non-rational play, i.e., whose advantage depends on the opponent putting weight on irrational or systematically mistaken behavior.

Similarly to formalizing suckerhood in the second step of the proof, formalize "hustling is possible at ω" as:

$$Hustle_i(\omega) \ := \ \exists j \neq i \text{ such that } a_j(\omega) \notin \mathrm{Rat}_j(G)$$

basically conveying that exploitation indeed requires some non-rationalizable opponent action/belief support. Returning to the first step of the proof it became clear that at any $\omega \in CK(R)$ we have $a_j(\omega) \in \mathrm{Rat}_j(G)$ for all j. So

$$Hustle_i(\omega)$$

is *false* for all i. There is no non-rationalizable "mark" to exploit. Hence, under common knowledge of rationality hustlers disappear because their defining fuel—others' systematic epistemic mistakes—cannot occur.

QED

Under common knowledge of rationality, the epistemic environment that creates the hustler/sucker asymmetry collapses. That's the act of disappearance. Everyone's realized actions must be rationalizable. Therefore suckers (players choosing non-rationalizable, exploitable actions due to mistaken beliefs) cannot occur. So hustlers (players whose gain relies on those mistakes) cannot occur either.

3.3 Equilibrium without Exploitation

What replaces the hustler/sucker distinction under common knowledge of rationality is not moral harmony but *strategic symmetry*. Agents may still disagree, compete, and pursue conflicting interests. But their interactions will be governed by equilibrium concepts that exclude systematic exploitation based on belief errors. Any remaining asymmetries must arise from differences in payoffs, resources, or institutional constraints—not from predictable mistakes in reasoning. This does not mean that outcomes become fair or egalitarian. It means only that exploitation based on epistemic vulnerability disappears. Strategic advantage must be earned within the rules of the game, not extracted from others' misperceptions.

The disappearance of hustlers and suckers under common knowledge of rationality also explains why the distinction is so pervasive in real life. Common knowledge is extraordinarily difficult to achieve in large, complex, and mediated societies. On social platforms by way of example, scale, anonymity, and information overload prevent agents from knowing what others know, let alone knowing that others know that they know. Even when individuals are highly informed, they cannot safely assume that their knowledge is shared. As a result, belief asymmetries persist, and with them the hustler/sucker distinction. Exploitation survives because the epistemic conditions required to eliminate it are rarely met. Remove the asymmetry, and the distinction dissolves. In a world without common knowledge, however, hustlers and suckers are not anomalies. They are equilibrium phenomena.

3.4 "Agreeing to Disagree"

One of the most celebrated results in epistemic game theory is Nobel Prize Laureate in Economics, Robert Aumann's agreeing-to-disagree theorem (Aumann 1976). It turns out that this theorem identifies the precise epistemic failure that makes hustlers and suckers possible.

The original and very clever theorem conveys that if two Bayesian agents have a common prior, update by Bayes' rule, and their posterior beliefs are common knowledge, then their posteriors must be equal. They cannot "agree to disagree." Formally, if

- P is a common prior,
- $P_i\,(E \mid \Pi_i(\omega))$ and $P_j(E \mid \Pi_j(\omega))$ are posteriors,
- and these posteriors are common knowledge at ω,

then

$$P_i(E \mid \Pi_i(\omega)) = P_j(E \mid \Pi_j(\omega)).$$

Aumann's theorem is not primarily about disagreement. It is about the epistemic power of common knowledge. Its deep lesson is that persistent belief asymmetries require a failure of common knowledge under common priors.

The hustler/sucker distinction maps directly onto Aumann's framework. The defining feature is that a sucker holds systematically false or overly optimistic beliefs about others' behavior (e.g., trust, cooperation, honesty) and acts rationally conditional on those beliefs. In Aumann's terms the sucker's posterior differs from the hustler's, but the sucker *does not know* that the hustler's posterior differs, or does not know that the hustler knows that the sucker's posterior differs, etc. Without this epistemic opacity, the sucker could not persist.

On the other hand, the hustler's defining feature is that he/she anticipates others' belief errors and chooses strategies optimal given those errors. Translated into Aumann's realm of thought the hustler knows (or assigns high probability) that posteriors differ, but very importantly, this fact is not common knowledge. If belief asymmetry became common knowledge, the hustler's advantage would collapse.

Now suppose hustlers and suckers have common priors. Then combine the result demonstrated earlier (1) under common knowledge of rationality, all actions are rationalizable with Aumann's theorem (2) under common knowledge of posteriors, belief asymmetries disappear. Together, they imply **no sucker** because the sucker's defining feature is a belief that is *both false and exploitable.* Aumann shows that such a belief cannot persist once posteriors are common knowledge. Any belief difference would be forced to converge. Theorem (1) and (2) likewise imply **no hustler** as the hustler's payoff advantage relies on asymmetric beliefs. Aumann shows that common knowledge collapses belief asymmetry. With symmetric posteriors, there is no epistemic leverage to exploit. Hence agreeing-to-disagree is the epistemic space in which hustlers and suckers live. When that space is eliminated, so are they.

The core connection is that a hustler/sucker interaction is a structured violation of Aumann's conditions. Either common priors fails, Bayesian updating fails or common knowledge of posteriors fails. Most often priors differ *implicitly* (social norms, trust heuristics) and belief differences are not themselves common knowledge. This allows the sucker to remain sincerely optimistic while the hustler remains strategically cynical.

3.5 The Evil Cousin

The notion of pluralistic ignorance originates in social psychology. The phenomenon occurs when individuals privately reject a belief or norm but assume—incorrectly—that others accept it (Krech & Crutchfield 1948). Classic examples include students who privately oppose heavy drinking but believe others endorse it (Prentice & Miller 1993), citizens who privately doubt propaganda but believe it convinces everyone else (Bergmann 2010) or board members believing that everyone else is in favor of the new business strategy although no one is for real (Halbesleben & Buckley 2004).

Pluralistic ignorance is an *evil cousin* of common knowledge (Hendricks & Rendsvig 2016). To see this arrange the two notions side by side, starting with common knowledge:

> A group G has common knowledge of a proposition A if everybody in G knows that everybody in G knows that ... everybody in G knows that A, for *all* higher-order levels.

Where (common) knowledge is truth tracking and may make groups act as one, pluralistic ignorance utilizes that groups are composed of individuals that seldom know each others' exact thoughts to create highly inefficient groups. Pluralistic ignorance—like common knowledge—is a higher-order notion, but defined using belief (Kretch & Cruthfield 1948), (Hansen, Hendricks & Rendsvig 2013, Hendricks & Rendsvig 2016, (Bjerring et al. 2013):

> A group G is in a state of pluralistic ignorance with respect to proposition A if

1. All members of the group believe *not* A,
2. All members of group G mistakenly believes that everybody else believes A.

Pluralistic ignorance is unfortunately far easier to achieve than common knowledge and plays tricks on us regularly. In some cases, the result is harmless. In attempting to accommodate the preferences of the other, dating couples might go see one movie, where both would have preferred the same other. In other cases, the result may cause liver deficiency in long run/worst case scenarios: If everybody on campus seemingly enjoys heavy alcohol consumption, new students may attempt to fit in by following suit given social proof (Bicchieri 2017, Cialdini 2007, Prentice & Miller, 1993).

The key feature of pluralistic ignorance is *epistemic isolation*. Each agent has private belief but lacks confirmation that this belief is shared. As a result, behavior conforms to a perceived norm rather than to private belief. From the perspective of rationality, this behavior is not irrational. Acting on private belief alone can be costly if others do not act similarly. Under pluralistic ignorance, conformity is often the *dominant response*, even when the norm itself is privately rejected.

Pluralistic ignorance therefore explains how false norms persist. But it does not yet explain why some agents profit systematically from others' conformity. For that, the hustler/sucker architecture is needed. Here the sucker is the agent who acts *as if* a norm is shared, because acting otherwise without coordination would be costly. Pluralistic ignorance places agents precisely in this position. The sucker may privately doubt the value of an asset, doubt whether waiting in line at the fancy coffee-house or club is worth the wait, doubt the sincerity of political speech, the fairness of a platform or the legitimacy of a social demand. Yet again, they continue to act *as if* these were widely endorsed. Their action sustains the norm despite their disbelief. The sucker bears the cost of maintaining coordination under epistemic opacity and their vulnerability is structural rather than cognitive.

The hustler occupies a different epistemic position. The hustler suspects or knows that pluralistic ignorance exists, knows that others privately doubt what they publicly enact, and exploits the fact that this doubt is not common knowledge. Hustlers need only ensure that others believe others believe. They profit from the gap between private skepticism and public conformity. In markets, hustlers exploit pluralistic ignorance by sustaining participation in bubbles long after skepticism is widespread. In politics, they exploit it by mobilizing performative support rather than genuine belief. On platforms, they exploit it by extracting engagement from users who privately resent the system but continue to participate.

Pluralistic ignorance is often described as a mismatch between private belief and public behavior. But the deeper issue is again non–common knowledge. Common knowledge is not merely shared belief. It is shared belief *known to be shared*, and known to be known, and so on. For coordinated action to occur—withdrawal, resistance, refusal—it is not enough that individuals privately agree. They must know that others agree, and know that others know that they agree.

Pluralistic ignorance is precisely the failure of this condition. Individuals may share doubt or skepticism, but because this skepticism is not publicly affirmed or mutually acknowledged, it cannot guide action. This is why pluralistic ignorance is so resilient. It is not dispelled by information alone. It requires actual *public epistemic shifts*—moments when private beliefs become mutually visible usually through some kind of public announcement.

When common knowledge emerges, this asymmetry collapses then sucker strategies cease to be rationalizable. Acting *as if* the norm holds is no longer a best response. At the same time, hustler strategies lose their foundation and potency, because exploitation depends on others continuing to play along. This is why moments of common knowledge—public scandals, sudden crashes, collective revelations—often produce rapid *behavioral* shifts.

3.6 Hustlers, Suckers and Behavioral Economics

Behavioral economics (Kahneman & Tversky 1979, Kahneman 2011, Thaler & Sunstein 2008, Shiller 2017) is often invoked to explain why some agents appear systematically vulnerable to exploitation. Cognitive biases, heuristics, framing effects, and bounded rationality seem to offer ready-made explanations for suckerhood, while hustlers appear as agents who understand or exploit these biases. From this perspective, the hustler/sucker distinction looks like a familiar asymmetry between sophisticated and unsophisticated decision-makers.

At first glance, sucker behavior maps neatly onto well-documented cognitive biases. Overconfidence, optimism bias, availability heuristics, social proof, and loss aversion all seem relevant. Agents may underestimate downside risk, overweight salient narratives, follow others' behavior, or delay exit to avoid realizing losses. Behavioral economics shows convincingly that real agents do not behave like perfectly rational optimizers. The sucker is the biased agent who misjudges probabilities, succumbs to emotional framing, or fails to update beliefs appropriately. The hustler, by contrast, appears as the debiased actor—one who understands how others systematically err and designs strategies to exploit these predictable deviations. But many hustling strategies do, in fact, exploit known behavioral regularities. Marketing, financial speculation, platform design and casinos rely heavily on nudges, salience manipulation, and framing effects (Thaler & Sunstein 2008, Schüll 2012). Thus, behavioral economics indeed provides a powerful toolkit for analyzing micro-mechanisms of exploitation.

There is one troubling issue nevertheless. The hustler/sucker distinction persists even when agents are aware of the biases. In financial markets, on social platforms, casinos and with regard to political communication participants openly acknowledge manipulation, cognitive limitations, dark patterns and emotional exploitation. They may even joke about them. Yet participation continues. This persistence cannot be explained by bias alone.

If suckerhood was simply a matter of misperception or error, learning and reflection should gradually eliminate it. Behavioral economics itself predicts debiasing effects over time (Gigerenzer 2018). But exploitation often survives widespread awareness. The reason is that suckerhood is not merely an individual cognitive failure. It is a strategic position sustained by epistemic uncertainty about others' behavior. An agent may for sure fully understand the biases at play and still find it rational to participate. The reason being that the payoff depends not on what he/she believes, but on what the agent in question believes *others* will do.

Here the behavioral perspective may be supplemented

by the game theoretical ditto. Many actions which look biased from an individual perspective are strategically rational responses to uncertainty about coordination. Consider herd behavior (Banerjee 1992, Bikhchandani, Hirshleifer, D. & Welch, 1992). Behavioral economics explains herding through social proof or conformity bias. But in many market contexts, herding is rational: If prices depend on others' actions, following the crowd may maximize expected payoff even when one privately disagrees with the crowd's beliefs. The agents are responding to incentives created by others' expected behavior.

Similarly, loss aversion explains delayed exit in declining markets (Shiller 2000, Kahneman 2011). But in coordination games, waiting may be rational if one expects others to continue holding. The suckers don't cling irrationally to hope but respond to uncertainty about timing. Behavioral biases thus *interact with strategic reasoning.* They shape beliefs, but do not fully determine behavior (Gigerenzer 2008). Hustlers exploit biases, but they also exploit the fact that belief correction does not automatically enable unilateral exit.

For the full story, the trick may be to combine behavioral economics with the role of non–common knowledge. Agents may know about biases, manipulation, and risks. What they do not know is whether others are ready to act on that knowledge. Without this higher-order certainty, coordination fails. Behavioral economics rarely addresses this epistemic layer because it, reasonably enough, focuses on individual cognition rather than collective belief structures (Gennaioli, Shleifer & Vishn 2018). Now, the hustler/sucker situation persists precisely because collective rationality is unavailable even when individual rationality improves. And you may be a hustler one minute, yet a sucker the next. In such settings, exploitation does not alone depend on deep cognitive error alone. It depends on delayed alignment of beliefs about beliefs. Biases may trigger initial dynamics, but epistemic fragmentation sustains them.

Behavioral economics provides indispensable insight into the micro-foundations of the hustler/sucker distinction. It explains how framing, emotion, and heuristics shape perception and vulnerability. It clarifies why exploitation is psychologically effective and how hustlers design strategies that resonate with human cognition. But it does not, in and by itself, explain why the distinction persists among informed, reflective agents, nor why insight so often fails to liberate. For that—is the current suggestion—consult the epistemic and game-theoretic perspective that treats suckerhood not as a mistake, but as a structural role in coordination under uncertainty. Seen this way, behavioral economics diagnoses the symptoms of suckerhood, but not its tragic cause.

The tragedy of modern agency lies not in being biased, but in acting rationally within systems that make vulnerability unavoidable and coordination perpetually out of reach. That may also account for how some financial bubbles come about.

CHAPTER 4: Hustlers, Suckers and Bubbles

Financial bubbles pose a perennial puzzle. They occur repeatedly, across eras, markets, and asset classes. They involve participants who are often highly educated, well-informed, and professionally sophisticated. And yet they display patterns that appear—from the outside— irrational as prices detach from fundamentals, skepticism coexists with exuberance, and collapse is widely anticipated yet insufficient to prevent participation. Hustlers and suckers to the explanatory rescue.

4.1 Bubble Beasts of Beliefs

In financial bubble studies, no model fits all (Kindlerberger & Aliber 2005, Vogel 2009, McCarty et al. 2013). Some explanations emphasize naïveté. Investors are misled, fooled, or cognitively biased. On the other end of the explanatory spectrum lies malfeasance. Manipulators deceive, exploit, or engineer fraud. For sure, both accounts capture something real, but neither explains the structural stability of

V. F. Hendricks, *Sucker Nation*,
https://doi.org/10.1007/978-3-032-22788-1_4

financial bubbles—why they persist even when deception is suspected, criticism is widespread, and intelligence is abundant. Financial bubbles may be cast as large-scale hustler/sucker equilibria sustained by failures of common knowledge not failure of intelligence.

To get the idea off the ground, step one is to abandon the idea that "suckers" in bubbles are just plain stupid (Abreu & Brunnemeier 2003). In market contexts, suckerhood rarely takes the form of simple false belief. Most participants in bubbles know prices may be inflated, know others suspect the same and know that manipulation is possible or even likely. What distinguishes suckers from hustlers is not belief in fundamentals, but belief about others' beliefs. In turn, the sucker is the agent who behaves *as if* enough others believe in the asset's future appreciation, participation will continue long enough to justify entry and exit will be possible before collapse. There are greater fools out there that will take the beating—and I, myself, have a chair when the music stops (van Lee 2017, Zhang & Zheng 2017). Now, the hustler is the agent who understands that belief itself, not value, is the price driver, that others' participation can be anticipated and exploited and one may profit by entering early and exiting strategically. These are not psychological type castings. Hustlers and suckers are but strategic positions generated by the epistemic structure of the market.

Revisit Aumann's central insight from the preceding chapters. Disagreement or divergence persists only in the absence of common knowledge. When rationality is common knowledge, certain forms of persistent misalignment disappear. Agents rule out systematically mistaken expectations about others' behavior. But financial bubbles are precisely environments in which rationality is not common knowledge. Participants may be rational. They may know others are rational. But they do not know whether others know that others are rational, whether skepticism is widely shared or privately held nor do they know whether withdrawal will be coordinated or isolated. This higher-order uncertainty is again crucial. It prevents agents from ruling out exploitable behavior by others, and it prevents coordinated exit even when collapse is anticipated.

In such environments, rationality coexists with exploitation. Game theory sharpens this point through the concept of rationalizability. To rehearse, an action is rationalizable if it can be justified as a best response to some belief about others' actions, assuming they are rational playing rationalizable actions. In bubble markets buying an overvalued asset can be rationalizable if one believes others will continue buying, holding despite skepticism can be rationalizable if one believes exit will be staggered and entering late can be rationalizable if one believes price momentum will persist. They are best responses to beliefs about others' beliefs. Without common knowledge of rationality, agents cannot eliminate these strategies through iterated reasoning. Sucker strategies remain rationalizable, and therefore persist. Hustler strategies—front-running, narrative amplification, strategic exit—are rationalizable *because* sucker strategies persist. The equilibrium survives not because participants are mistaken, but because *no one can rule out others' mistakes.*

A hustler/sucker equilibrium in a financial bubble does not require that suckers believe the asset is fundamentally sound. It requires only that they believe others will continue to act *as if* it is long enough to justify participation. This is why bubbles often involve a measure of cynical participation. Investors openly acknowledge overvaluation while continuing to buy. The familiar justification—"I know it's a bubble, but I can get out in time"—is not cognitive dissonance. It is equilibrium reasoning under epistemic uncertainty.

Each agent's strategy is a best response to expected behavior. No unilateral deviation improves outcomes as exiting early risks foregone gains if the bubble persists while staying exposes one to crash risk if others exit first. So, the equilibrium holds because exit is individually costly without coordination. Within this equilibrium, hustlers occupy a distinctive role. Hustlers are not merely manipulators in the crude sense. They are *epistemic arbitrageurs* meaning agents who profit from the gap between belief and belief-about-belief. They exploit delayed realization of common skepticism, informational asymmetries about timing and narrative momentum rather than fundamentals (Shiller 2017). As before, hustlers only need to maintain uncertainty about when others will stop believing. Even public skepticism can serve this purpose by signaling sophistication while failing to coordinate exit. In this sense, market manipulation often consists not in downright lying, but in managing higher-order beliefs signaling confidence, amplifying narratives and obscuring exit intentions. The sucker's vulnerability lies not in believing the story, but in not knowing when belief will cease to be operative.

4.2 Coordination Failure

The tragic core of bubbles lies in coordination failure. Many participants would prefer a world in which bubbles did not inflate, manipulation was ineffective and prices actually tracked fundamentals. Yet given the epistemic structure, unilateral restraint is dominated. Refusing to participate does not discipline the market. It merely excludes the refuser from gains while others continue. The sucker remains rational precisely because collective rationality is unavailable. Aumann's framework clarifies why. Without common knowledge that others will also withdraw, no agent can rationally withdraw first. Each waits for signals of coordinated exit that never arrive in time. The result is a stable but fragile equilibrium—well, stable until it suddenly collapses.

When bubbles burst, the collapse often appears abrupt, even when warning signs were abundant (Lewis 2011). This is not mysterious. It reflects an *epistemic phase transition*: A sudden shift in higher-order beliefs. At some point, participants come to believe not merely that the asset is overvalued, but also that others now believe this and *know* that others believe it. In other words, skepticism approaches common knowledge. At this point, sucker strategies cease to be rationalizable. Holding or buying can no longer be justified as best responses. Exit cascades. The tragedy is that this realization typically arrives *too late* for many participants. Insight does not save them—it precipitates collapse.

From this perspective, market manipulation is not an external distortion of an otherwise efficient system. It is *endogenous to the epistemic structure* of markets without common knowledge. Manipulators succeed not by overpowering rationality, but by *operating within its limits.* They exploit the fact that belief alignment is costly, coordination is fragile and timing is asymmetric. This explains why manipulation can be widespread even in heavily regulated, information-rich markets. Transparency alone does not generate common knowledge. Public information does not translate automatically into coordinated action.

Now there is a temptation to moralize bubbles by casting hustlers as villains and suckers as victims. This temptation should be resisted. Hustlers respond rationally to incentives created by belief asymmetries. Suckers respond rationally to coordination constraints. Both act within a system that requires some agents to bear vulnerability for others to extract profit. The tragedy is not moral failure. It is structural incompatibility between individual rationality, collective rationality and epistemic transparency. No amount of intelligence or education abolishes this tension.

Financial bubbles exemplify the tragic structure of modern agency at scale. They show how rational agents, acting intelligently and reflexively, may nonetheless become trapped in equilibria that expose some to loss and others to gain. The hustler/sucker distinction, grounded in Aumann-style epistemic reasoning, reveals that bubbles persist because no one can safely stop playing alone.

Financial markets are tragic games. They require trust in expectations that are exploitable, participation in dynamics that one knows are unstable, and action under conditions where insight does not guarantee escape. Bubbles end not when truth is revealed, but when belief becomes common knowledge—and by then, it is already too late. That is not a pathology of markets. It is the price of coordination without common knowledge.

4.3 Crypto and Meme Stocks

Crypto markets and meme stocks are often discussed as anomalies—spectacles of irrational exuberance, youth culture, make energy, internet hype, or mass delusion (Arash, Ouzan, & Shahzad 2022, Shi 2022, Martin et al. 2025). Crypto is framed as technological utopianism run amok; meme stocks as collective trolling mistaken for finance. Both interpretations comfort traditional market theory by treating these episodes as deviations from normality.

On the current reading crypto and meme stocks are not deviations. They are exemplary cases of the hustler/sucker equilibria in the Aumann-style epistemic analysis. Crypto and meme stocks alike are revealing cases of how rational agents can participate knowingly in bubbles, how exploitation can persist without deception, and how market dynamics can stabilize around belief asymmetries rather than fundamentals. They are therefore ideal sites for applying the epistemic hustler/sucker framework and Aumann-style epistemic analysis.

At the center of both crypto and meme-stock dynamics again lies the, by now, familiar epistemological notion of beliefs about beliefs. Participants rarely need to believe that a crypto token has intrinsic value, or a meme stock is worth its price. As before, they only need to believe that *others* believe—or will continue to act *as if* they believe long enough to justify participation. The distinction is decisive because it shifts analysis away from truth, value, or fundamentals, and toward higher-order expectations or the lack of common knowledge. In both crypto and meme stocks skepticism is widespread, irony is explicit and critique is public. Yet participation persists. This persistence cannot be explained by false belief. It must, as before, be explained by epistemic uncertainty about coordination.

The hustler/sucker distinction scales cleanly to both cases because suckers are not fools but agents who sustain participation under the belief that enough others will continue to participate. And hustlers are agents who exploit this belief asymmetry. The same individual may occupy different positions over time, sucker one minute, hustler the next. What matters is not who someone is, but how their strategy relates to others' beliefs. In both crypto and meme stocks, suckerhood is often openly acknowledged. Participants joke about "being exit liquidity" (Crypto.com

2026). This irony does not dissolve suckerhood—it stabilizes it. Irony signals sophistication without enabling coordination.

Crypto stocks are shares of publicly traded companies whose business models are heavily tied to cryptocurrencies, such as crypto exchanges, mining firms, or companies holding large amounts of digital assets. Their value often fluctuates with crypto market sentiment and prices, making them more volatile than traditional stocks and closely linked to developments in the digital asset ecosystem. Thus, crypto assets represent an extreme case of expectational value. Unlike equities, they lack cash flows, lack institutional valuation anchors and lack widely accepted use-value benchmarks (Urquhart & Yarovaya 2024). Their price depends almost entirely on expectations of future expectations.

This makes crypto markets epistemically transparent. There is little pretense that prices reflect fundamentals. Instead, prices reflect network effects, the narrative momentum, belief in adoption and ... belief in belief. Crypto thus operates almost entirely in the space Aumann identified as disagreement and coordination persist due to non–common knowledge.

Crypto hustlers may include early adopters, influencers, founders and insiders with timing advantages. Their strategy does not depend on convincing others of intrinsic value. As with financial bubbles, it depends on managing higher-order beliefs amounting to sustaining uncertainty about when others will stop believing, signaling confidence without committing to long-term holding and amplifying narratives while retaining exit optionality. Like in the case of financial bubbles, crypto hustlers are epistemic arbitrageurs. They profit from the lag between widespread skepticism and coordinated withdrawal.

Crypto suckers are not necessarily believers in utopia. Many explicitly acknowledge volatility, manipulation and lack of fundamentals. Reciting the case of financial bubbles, suckers again participate because exit is costly without coordination, abstention does not discipline the market and timing dominates valuation. They act rationally under epistemic opacity. Holding or buying remains rationalizable as long as others might continue.

Crypto bubbles persist because belief alignment is structurally impossible, transparency does not produce common knowledge and irony substitutes for coordination. Even public skepticism fails to discipline prices, because it does not resolve uncertainty about *when* belief will collapse. Thus Crypto is not irrational finance. It is epistemic finance in pure form.

Now, if crypto represents *radical opacity*, meme stocks represent *epistemic inversion*. Meme stocks are shares of companies whose prices surge primarily due to viral attention and coordinated buying by online communities rather than changes in underlying fundamentals. Their volatility is driven by social media momentum, short squeezes, and collective speculation, making price movements highly sentiment-dependent. By way of example, meme stocks like GameStop involved open acknowledgment of overvaluation, explicit rejection of fundamentals and performative participation (Cao & Liu 2022). Participants often stated outright that the stock was overpriced. This did not deter participation. It reframed it. The key epistemic move was not belief, but coordination around belief-about-belief: "It doesn't matter what I believe, as long as others believe that others believe."

Meme-stock hustlers may include agents like early entrants, options traders, narrative amplifiers and opportunistic funds exploiting volatility. Unlike crypto, hustlers here often masqueraded as anti-hustlers—claiming to act for justice, revenge, or community. This moral overlay did not negate the epistemic structure but merely it intensified it. The hustle was not hidden. It was *collectively disavowed while individually practiced.* Pluralistic ignorance of the normative kind.

Suckers in meme stocks often embrace labels like "We are all going to lose money" as a badge of solidarity (Jakob 2026). This self-awareness did not dissolve suckerhood. It stabilized it by deflecting shame, masking strategic asymmetries and postponing coordination failure. Irony again played the same role as before. That of signaling sophistication without enabling exit.

Despite their cultural novelty, meme stocks conform exactly to Aumann's insight. Persistent participation despite disagreement signals lack of common knowledge. Everyone may know the stock is overpriced. Everyone may know that everyone knows this. What they do not know is when others will act on that knowledge. Thus best response persists. Holding remains a best response under uncertainty.

The real difference between crypto and meme stocks lies in how epistemic non–common knowledge is expressed. Where crypto emphasizes items like futurity, innovation, and inevitability, meme stocks emphasize irony, antagonism and performativity. Crypto promises meaning later. Meme stocks deny meaning now. Yet both converge structurally.

In both cases, collapse occurs not when truth is revealed, but when belief alignment shifts, skepticism approaches common knowledge and sucker strategies cease to be rationalizable. Collapse is sudden because epistemic shifts are nonlinear. Once agents believe that others will exit, exit becomes rational for all. The tragedy is the same: Realization arrives too late for many participants. Insight does not save them but merely synchronizes loss.

With crypto and meme stocks, manipulation should be understood as *epistemic engineering*. Manipulators succeed by sustaining uncertainty, delaying common knowledge and exploiting coordination failure. This explains why regulation and disclosure alone are insufficient. *Information alone does not produce common knowledge.*

4.4 Go Long. Go Short. Just Go!

Crypto and meme stocks may be understood as different manifestations of the same underlying hustler/sucker equilibrium under epistemic non–common knowledge. But the same goes for what in economics is referred to as *day trading* (Douglas 1990, Murphy 1998) independently of whether crypto or meme stocks or some other asset class is peddled that day. All three abandon fundamentals in favor of short-horizon expectations about others' behavior. Price movements are driven not by value, but by beliefs about belief. Crypto radicalizes this logic by operating almost entirely on expectational value while meme stocks theatricalize it by openly rejecting valuation while sustaining coordination through irony and identity. Day trading operationalizes it as continuous action, whether long or short.

Unlike long-term investing, day trading openly abandons narratives of fundamentals, value creation, or economic growth (Lefèvre 1923). Participants do not typically claim that prices reflect intrinsic worth. Instead, day trading is explicitly about anticipating short-term price movements driven by order flow, sentiment, and other traders' actions. This makes day trading an unusually honest financial practice. It admits upfront that markets are games of expectations (Lewis 2011).

Yet this honesty does not prevent exploitation. On the contrary, day trading repeatedly reproduces stable asymmetries in which a small minority profits while the majority loses. These losses persist even when participants are informed, reflective, and aware of the odds. The hustler/sucker framework may explain why. Day trading is not sustained by false belief about value. It is sustained by epistemic uncertainty about timing, coordination, and relative position. Whether one goes long or short, the same tragic structure applies.

Going long in day trading means buying an asset with the expectation of selling it at a higher price within a short time frame. The long trader does not need to believe the asset is fundamentally undervalued. They need only believe that others will buy after them—an idea reminiscent of greater fools theory in bubble formation. The long-side sucker is the one who believes they can enter early enough, ride momentum long enough, and exit before reversal. These beliefs are not obviously irrational. If enough traders behave similarly, momentum can be self-reinforcing. The sucker's vulnerability lies in *uncertainty about others' exit strategies*. Each trader may know that reversals happen. What they do not know is when others will decide to sell. Thus, the long trader's action is best response under non–common knowledge. Buying remains a best response as long as it is plausible that others will continue buying. Losses occur not because the trader was wrong about direction, but because they were late relative to others' exits.

Short selling introduces an apparent asymmetry. The short seller bets on price decline rather than increase. Superficially, this looks more sophisticated. Short sellers often pride themselves on skepticism, contrarianism, and resistance to hype. They pretty much claim to see through bubbles (Guimaraes & Pannella 2024). Yet short day trading reproduces the same sucker dynamics. The short-side sucker believes they can identify overextension, time reversals accurately and cover before momentum resumes. Here the vulnerability is even sharper. Prices can rise faster than they fall, and losses on short positions are theoretically unlimited. The short trader may be *correct* about overvaluation and still lose money due to timing and coordination failures.

The short seller may know an asset is overpriced. They may know others know this too. What they cannot know is how long others will continue to trade *as if* it were not. As long as uncertainty about others' beliefs persists, remaining short can be rationalizable—and ruinous. Skepticism or doubt does not protect against suckerhood. Going short replaces faith in growth with faith in collapse timing. Both rely on beliefs about others' beliefs.

Hustlers in day trading are rarely defined by directional bets. They are defined by *structural advantage.* Market makers, high-frequency traders, brokers, and algorithmic systems profit regardless of whether prices rise or fall (Lewis 2014). Their gains depend on volume, volatility, spreads and predictable retail behavior. From their perspective, both long and short retail traders function as suckers. Each trade provides liquidity and revenue. Direction is irrelevant. Participation is the asset.

Even directional hustlers—those who go long or short—differ from suckers primarily in *timing and exit control*, not belief accuracy. Hustlers enter earlier, they exit faster, manage risk asymmetrically, and externalize losses onto slower participants. Their advantage lies in *epistemic position*, not superior insight into value. One might expect that awareness of these asymmetries would eliminate suckerhood. In practice, it does not. Day traders frequently acknowledge that most traders lose, that platforms profit from volume and that timing is brutally difficult (Gallegos-Erazo 2024). This awareness often takes ironic form. Here traders joke too about being "exit liquidity" (Crypto.com 2026). They acknowledge manipulation while continuing to trade. Irony here plays a stabilizing role. It signals sophistication without enabling exit. This mirrors the broader hustler/sucker equilibrium. Awareness does not produce withdrawal because withdrawal is unilateral and costly. To stop trading is to forfeit the possibility of being among the winners. As long as participation continues, the game remains rationalizable.

Long and short positions also map onto different forms of *identity* and dissonance management. Long traders often frame themselves as optimistic, forward-looking, and adaptive. Short traders frame themselves as critical, analytical, and contrarian (Curtis & Fargher 2013). Both identities serve to manage cognitive dissonance when losses occur. Losses are reinterpreted as being early, being disciplined and learning experiences. Traders trade *wits for vanity*.

Strategic clarity about structural disadvantage is sacrificed in exchange for a self-concept that preserves agency and competence. Directional conviction becomes a psychological buffer against recognizing suckerhood. Hustlers benefit from these identities. Platforms and educators encourage both bullish and bearish narratives, because both generate engagement and trades—that's all what's wanted anyway, so why preach otherwise?

The epistemic hustler/sucker architecture shows that day trading is not sustained by directional error. Going long and going short are symmetric expressions of the same epistemic condition of *acting under uncertainty about others' timing*. Suckers are not those who bet the wrong way. They are those who must bet at all to remain agents. Hustlers are not those who predict better. They are those who profit regardless of direction. Day trading shows how rational, aware agents can persist in disadvantageous roles, how skepticism fails to protect, and how action itself becomes the commodity.

4.5 By the End of Trading Day

Crypto, meme stocks and day trading embolden the central tragic insight. Markets can be rational, reflexive, and ironic—and still systematically exploit some participants. Suckers are not naïve. Hustlers are not omniscient. Both are trapped in a structure where exit requires coordination that cannot be guaranteed. By the end of trading day, whether one goes long or short, the tragic structure remains the same: To act is to risk being a sucker yet refusing to act is to risk ceasing to be an agent at all. The hustler/sucker distinction reveals that modern markets may not be efficient machines, but tragic games, where wisdom arrives too late, innocence cannot be restored, and agency is exercised under permanent epistemic damage. That is not just a pathology to be cured. It is the condition under which modern finance now operates. It is also the condition under which attention economics operate.

CHAPTER 5: Hustlers, Suckers and Attention

A central thesis of attention economics is that modern societies no longer suffer from a lack of information but from a surfeit of it. The bottleneck is attention: What agents notice, process, trust, and act upon. Attention scarcity is not merely a cognitive constraint but a *systemic vulnerability.* When attention is scarce and information abundant, agents must rely on heuristics, signals, and social cues to decide what matters. This necessity opens the door to *epistemic manipulation.* The hustler/sucker divide but names the social roles that emerge under these conditions in attention economics.

5.1 Attention Hustling

Back in 1971—way before the internet, social platforms and the tech-industry—Nobel Price Laureate in Economics, Herbert Simon, said something very prophetic about an age with information in abundance: "In an information-rich world, the wealth of information means a dearth of

V. F. Hendricks, *Sucker Nation*,
https://doi.org/10.1007/978-3-032-22788-1_5

something else: A scarcity of whatever it is that information consumes. What information consumes is rather obvious: It consumes the attention of its recipients" (Simon 1971).

Attention is a scarce and valuable cognitive resource—but also a potentially highly profitable capital asset. Herbert Simons' insight is the basis on which the term "attention economy" is coined. In a market for information products ruled by an attention economy the trick is to come up with a business model in which one may profit from exchanging information for attention. The "attention economic business model" is behind some of the worlds' largest corporations and has gained outsized importance not only on people's daily lives but also on the conditions for democracy in the 21st century (Hendricks & Vestergaard 2019, Hendricks & Mehlsen 2022)

In line with Simon's insight, on social platforms, attention is scarce, monetizable, algorithmically allocated and strategically manipulated. Social platforms do not primarily sell content. Really what they sell is *access to attention flows* (Webster 2014, Wu 2016, Zuboff 2019) This transforms epistemic life into an economic field where visibility, virality and outrage function as currencies. But attention is not neutral. What captures attention shapes belief formation, salience, and perceived consensus. Thus, attention allocation is also belief allocation. This is where hustlers and suckers enter.

An *attention hustler* is an agent (individual or institutional) who understands how attention is captured (emotion, outrage, novelty) (Berger 2013, Aral 2020), how attention may be mistaken for importance or truth and that this mistake is not common knowledge. Now actors need not convince audiences of false propositions. It is enough to shift attention, distort salience or crowd out epistemic competitors (Hendricks & Vestergaard, 2019). Attention hustlers exploit the fact that what is most attended to is often taken to be what is most believed. This is clearly false—but may nevertheless be rational to assume under attention scarcity. The list of examples is extensive but to mention af few like outrage entrepreneurs and rage baiters, disinformation campaigns, clickbait media, platform influencers and political actors gaming visibility rather than truth. These actors are not epistemically confused. They are epistemically strategic.

By contrast, the *attention sucker* is not gullible in the traditional sense. They are agents who treat attention as a proxy for importance, visibility as a proxy for consensus, and repetition as a proxy for truth (the illusionary truth effect) (Vosoughi et al., 2018). Under attention scarcity, this behavior is individually rational. Agents cannot verify everything so they must rely on cues and heuristics. The sucker's epistemic error is a false inference about the epistemic environment: "If this commands attention, it must matter." Or "If everyone sees this, everyone must believe it." What the sucker does not know is how *strategically engineered* attention has become—or, more precisely, how widely this strategic engineering is understood.

This gives rise to epistemic pollution and non-common

knowledge. Epistemic pollution does not primarily consist in false information, but in *distorted informational environments* that systematically mislead attention and inference. This aligns directly with the hustler/sucker framework as:

- Hustlers profit from information pollution.
- Suckers suffer from it.
- The pollution persists because its effects are not common knowledge.

So even when individuals know that attention is manipulated, they do not know how much manipulation there is, who is manipulating, whether others recognize manipulation and whether others know that others recognize it. This higher-order opacity once again here sustains exploitation.

By way of example, there is a reason why outrage dynamics and rage baiting more often than not dominate digital attention economies. Ourtrage captures attention efficiently (Berger 2013), bypasses deliberation (encourages knee-jerk reactions), amplifies identity signaling ("I'm angry too just like you") and spreads regardless of truth value. From the hustler/sucker perspective outrage entrepreneurs are hustlers who convert moral attention into visibility and power, while participants are suckers insofar as they treat outrage participation as epistemic contribution rather than attention labor. The key asymmetry amounts to the fact that hustlers know outrage spreads independently of truth. Suckers believe outrage spreads in part because of truth or importance. Outrage dynamics produce a distinctive sucker position because agents believe they are participating epistemically (condemning falsehoods), but what they really are doing is supplying attention to epistemic pollutants—feeding the beast. The outrage hustlers understand this all too well. They provoke outrage intentionally, they know outrage spreads faster than correction. They know that participants will mistake moral reaction for epistemic success.

This is hustling at scale and besides the hustlers knowing this all too well, social platforms themselves roll along as meta-hustlers (Hendricks & Mehlsen, 2022). They optimize for engagement, not epistemic quality. They know that users conflate attention with value and know once again that this conflation is not common knowledge. Yet they frame attention extraction as a form of personal expression, a way of community building even democratic participation. This is a paradigmatic hustler move. What you are really profiting from is others' misinterpretation of the epistemic meaning of their own behavior. That's how results and insights from social psychology may be used in way detrimental to human agency and self-determination (Hendricks & Hansen 2016, Berger 2017).

5.2 Fake News for Suckers

Fake news is not simply false information (Hendricks & Vestergaard, 2019, Harris 2022, Broda & Strömbäck 2024).

What matters is the quality of the information as a mixture of truth and falsity and everything in between and the epistemic dysfunction of fake news (Hendricks & Hansen, 2016). That is distortions in how attention, trust, salience, and credibility are distributed. Fake news is best understood as a form of the aforementioned epistemic pollution disrupting the environment in which beliefs are formed, hijacking attention rather than persuading rationally while crowding out epistemically valuable content.

This perspective dovetails exactly with the distinction between hustlers and suckers. Fake news succeeds not because people are half-wits, but because they are epistemically positioned as suckers in an attention-scarce environment. Fake news actors often do not care whether audiences believe the content. What they care about is the attention allocation, the emotional arousal, the algorithmic amplification and agenda-setting.

In the attention economics setting, the fake news producer (state actor, clickbait entrepreneur, outrage broker or rage baiter) is a hustler in the precise sense developed. They understand the attention economy. They are well aware that audiences infer importance from visibility. They know this inference is not common knowledge and exploit it strategically. Fake news hustlers do not necessarily need to deceive about facts. They exploit beliefs about belief formation.

Conversely, the fake news sucker is not necessarily someone who believes the fake news content. More often, it is someone who shares it "just in case", reacts outrageously to debunk it, treats virality as evidence of relevance and again interprets repetition as endorsement. So even critical engagement can be epistemically damaging, because it feeds the attention economy. As before, the sucker's error is higher-order problem. Fake news suckers essentially misjudge what attention does in a polluted epistemic environment.

Fake news thrives under attention scarcity (Hendricks & Vestergaard 2019). Because agents cannot verify everything on the way and in their (news)feed they must rely on heuristics like popularity, emotional resonance, social endorsement and familiarity. While these heuristics may be rational under normal conditions they become exploitable under strategic manipulation. This precisely amounts to the structural condition generating suckers acting rationally given limited attention while being systematically exploitable by those who understand the system. Hustlers, by contrast, are agents who understand the very heuristics and their limits and very importantly understand that others must rely on them.

The real danger of fake news is *degradation of epistemic trust*. Fake new blurs distinctions between credible and non-credible sources, overloads attention and makes epistemic vigilance too costly. From the hustler/sucker point of view fake news is not about making people believe X, but rather about making people unsure *who to trust*, and unsure whether others know who to trust. This uncertainty creates fertile ground for hustlers because strategic actors thrive when epistemic standards collapse, and because belief asymmetries become persistent and non-common knowledge.

Fake news ecosystems normalize *persistent disagreement.* This is where Aumann's agreeing-to-disagree theorem again becomes relevant. Aumann shows that persistent disagreement requires a failure of common knowledge. Fake news ecosystems are engineered precisely to prevent belief convergence, prevent shared epistemic baselines and finally prevent common knowledge of credibility. Thus fake news is the *industrialization of agreeing-to-disagree.* And while hustlers exploit this condition knowingly, suckers inhabit it unknowingly.

A key normative lesson is that exhortations to "be critical" are insufficient. This lesson maps indeed cleanly onto the hustler/sucker distinction. The reason is that individual critical awareness in and by itself does not generate common knowledge. Add on top of this that knowing that fake news exists is not knowing who knows, nor knowing that others know that you know. This amounts to an additional reason why fake news is resilient to fact-checking, debunking, and education alone (Berger et al. 2025). Besides, these days social platforms are largely dropping fact-checking anyway (Brager 2025), it's not conducive to their business model and adds too much friction even if such friction may actually prove to being beneficial speedbumps to the proliferation of misinformation (Jahn, Rendsvig, Flammini, Menczer, & Hendricks 2025).

The fake news problem is not primarily about truth versus falsity, but about epistemic environment design.

The hustler/sucker separation provides a complementary lens that clarifies *why* this design is so effective. Fake news does not need fools. It needs attention scarcity, epistemic opacity, and trust under pressure. Fake news simply works because it turns ordinary, rational participants into structural suckers—while rewarding those who understand that this is happening. This goes all the way up to conspiracy theories as forms of fake news items on steroids (Hendricks & Vastergaard 2019).

5.3 Suckers for Conspiracies

Conspiracy theories are typically explained as pathologies of belief like products of paranoia, cognitive bias, low information, or distrust in institutions (Sunstein 2009b, Douglas et al. 2017, Prooijen & Douglas 2018). From this stance, conspiracy believers appear as epistemic failures—people who mistake coincidence for coordination or fiction for fact. But perhaps conspiracy theories persist not merely because some people believe false things, but because they offer a structural response to the hustler/sucker condition of modern society. They reorganize epistemic asymmetry by promising an escape from suckerhood—an escape that ultimately reproduces it. Conspiracy theory is not primarily about belief in secret plots. It is about *refusing to be the last one fooled* in a world where exploitation is widely suspected but unevenly understood.

Modern agents live in environments saturated with manipulation in terms of advertising, political spin, platform optimization, misinformation, and strategic communication. Many people know this—"we know a lot" as Lewis noted in the introduction. The sense that "things are rigged" or "we're being played" is widespread and often justified. This creates a pervasive anxiety of suckerhood. The fear is not merely of being wrong, but of being *naïve*—of taking official narratives, expert claims, or institutional assurances at face value while others see through them. Conspiracy theories feed directly on this fear. The conspiracy theorist often begins from a correct intuition that power is opaque, incentives are hidden, and public discourse is not truth-aimed (Sunstein & Vermuele 2009). Where they differ from skeptics is not in distrust, but in how they *resolve epistemic uncertainty.*

Conspiracy theories offer a seductive promise of *epistemic reversal.* Instead of being the sucker—the one who believes surface narratives—the believer becomes the one who "knows what's really going on" (Hofstadter 1964). The world is no longer confusing or contingent; it is secretly ordered. In hustler/sucker terms, conspiracy thinking attempts to *switch roles.* The believer positions themselves as a hustler of knowledge: Someone who sees through appearances, uncovers hidden coordination, and refuses to be fooled. This explains a key feature of conspiracy discourse in its obsession with exposure, revelation, and awakening. To believe a conspiracy theory is to believe that one has escaped epistemic exploitation. Yet this escape is fragile.

Conspiracy ecosystems are great at generating their

own hustlers. These may include influencers, content creators, grifters, and political entrepreneurs who profit from sustained distrust. Importantly, these hustlers do not need to prove their claims. They need only to maintain uncertainty and reinforce the sense that official knowledge is unreliable. Conspiracy hustlers exploit the—by now—familiar asymmetry that the believer must continuously interpret, connect, and verify while the hustler only needs to suggest doubt and provide narrative coherence. Thus, conspiracy theories mirror other hustler/sucker equilibria. The sucker is not deceived by a single false claim, but trapped in a *belief system that continually demands engagement*. The more effort invested, the harder exit becomes.

A central feature of conspiracy thinking is epistemic isolation. Believers often claim that "most people are asleep," "the truth is suppressed," or "only a few see it" but "the truth is out there" just "do your own research". These claims reflect not merely arrogance, but a response to non–common knowledge. In environments where trust in institutions collapses but no shared alternative epistemic authority emerges, knowledge fragments. Individuals know that manipulation exists, but do not know what others know or trust. Conspiracy theories resolve this uncertainty by positing a hidden but unified explanation.

This resolution is epistemically appealing but socially costly. It replaces uncertainty with certainty at the price of coordination with broader epistemic communities. The believer escapes one form of suckerhood by entering another. Once committed, conspiracy beliefs are stabilized by cognitive dissonance. Evidence against the theory is reinterpreted as evidence of cover-up. If you disagree it is only because you are part of the conspiracy. No such thing as negative evidence and a false theory, only evidence come what may and you are free to interpret it as you see fit and whatever serves your purpose. Skepticism from outsiders confirms the believer's sense of epistemic superiority.

This dynamic mirrors the sucker's position more generally. Continued investment in terms of "doing your own research" becomes necessary to justify past investment. Exit would require admitting not just error, but having been played—by influencers, communities, or one's own need and feed for epistemic security. Thus, conspiracy belief persists not because believers are impervious to evidence, but because withdrawal threatens identity, dignity, even agency.

5.4 Political Communication

Fake news, political communication—often joined by conspiracy theories—are thick as thieves these days. Political actors are routinely accusing the opposition or the legacy media of either being fake news or proliferating such. Political communication is frequently analyzed as actions of persuasion. Actors attempt to convince citizens of propositions, values, or policy preferences. When persuasion fails or becomes manipulative, the explanation is often enough psychological. Voters are misinformed, biased, emotional, or inattentive. Democratic failure is considered a deficit of rationality—a "rationality gap" —among citizens and a surplus of strategic cunning among elites (Paskalev 2020, Landwehr & Schäfer 2025, Schneider 2025).

In a sucker nation, political communication is less about belief formation but again much about managing beliefs about beliefs. Political actors do not need citizens to believe particular claims. They need citizens to believe that others believe them, or that acting *as if* others believe them is strategically necessary. This shift places political communication squarely within the hustler/sucker setup because in the political contexts, the sucker is not the ignorant voter. The sucker is the agent who continues to participate *as if* democratic norms still function in their classical form. This includes assumptions such as political speech aims at persuasion rather than mobilization, factual correction matters, free speech is intact, sincerity is rewarded and participation meaningfully influences outcomes. These assumptions are not reflections of stupidity. They are normative preconditions of democracy itself. Without them, democratic agency once more collapses into cynicism or apathy. Citizens must act *as if* their speech counts, free speech exists, their vote matters, and political communication is oriented toward reasons.

The tragedy is that these assumptions are increasingly exploitable. Large-scale media systems, algorithmic amplification, and polarized publics convert participation into data, outrage, and symbolic positioning rather than deliberation. So much for deliberative democracy (Sunstein 2009a). Yet citizens cannot simply abandon participation without abandoning democratic agency itself. The sucker, then, is the citizen who continues to engage—arguing, fact-checking, voting, protesting—despite knowing that these actions are often absorbed into strategic communication cycles that do not reward sincerity or truth. Their vulnerability is not cognitive; it is *institutional.*

The political hustler is not just a liar or propagandist. Hustlers may include political consultants, populist leaders, outrage entrepreneurs, strategic media actors and increasingly, platform-optimized communicators and influencers. What distinguishes hustlers is their relation to expectations. Hustlers understand that political communication functions less as deliberation than as *signal coordination.* They exploit attention dynamics, moral polarization, and identity signaling to mobilize support without requiring belief coherence. Political hustlers do not need their audience to believe their claims. They need them to repeat them, react to them, defend them as identity markers (Marker & Hendricks 2019), or believe that others believe them. Political hustling is once more a form of *epistemic arbitrage* similar to crypto and meme stock hustlers. Hustling politics profits from gaps between private skepticism and public performance, between knowledge and coordination. Irony, cynicism, and even explicit acknowledgment of manipulation often strengthen rather than weaken this strategy.

The persistence of manipulative political communication cannot be explained by deception alone. As already established, many citizens recognize spin, exaggeration, and bad faith. Yet participation continues. This persistence reflects a *stable equilibrium*. From the citizen's perspective withdrawing from political communication risks irrelevance, disengagement cedes the field to more extreme actors and unilateral refusal does not improve collective outcomes. From the hustler's perspective as long as participation continues, exploitation remains profitable, outrage and polarization sustain visibility while sincerity by others can be predictably instrumentalized. Neither side has a unilateral incentive to change. Citizens continue to engage despite frustration; hustlers continue to manipulate because it works. The equilibrium persists not because citizens are fooled, but because *coordination around restraint is unavailable*.

This once again mirrors Aumann-style epistemic failure. Citizens may know that political discourse is degraded. They may know that others know this. What they do not know is whether others are ready to withdraw, slow down, or refuse certain communication games. Without common knowledge, sucker strategies remain rationalizable.

Polarization (Sunstein 2009) intensifies the dynamics of among hustlers and suckers. In polarized environments, political communication becomes less about persuasion across divides and more about *mobilization within camps*. This further shifts communication from belief to belief-about-belief. Citizens may privately doubt extreme claims but publicly endorse them to signal loyalty in spirit of blue lies. They may share content they know is misleading because it functions as identity affirmation. Hustlers exploit this by crafting messages that are *epistemically thin but socially potent.* Non–common knowledge is central once more. Within groups, skepticism may be widespread but privately held. Public expression exaggerates certainty. Hustlers benefit from this divergence between private belief and public signal. Efforts at correction often backfire, reinforcing group identity rather than changing minds. The sucker, in this context, is not the partisan believer but the agent who continues to argue as if reasons matter symmetrically across camps.

A further stabilizing factor is the moralization of political participation. Citizens are told that silence is complicity, that engagement is a civic duty, and that withdrawal is irresponsible. These injunctions are not wrong. Democracy requires participation. Yet they also increase vulnerability. When participation becomes a moral obligation rather than a strategic choice, agents are pressured to expose themselves continuously to manipulative communication environments. Hustlers exploit this moral pressure by framing every issue as urgent, existential, and identity-defining. The sucker here is not merely misled; they are *ethically compelled* to participate under conditions that systematically convert participation into leverage for others.

Contemporary political communication operates within attention economies that reward speed, emotional salience, and conflict (Harsin 2015). These conditions favor hustlers who can provoke outrage, rage baiting, dominate cycles, frame opponents reactively and extract attention regardless of outcome. Citizens who respond earnestly—by correcting facts, offering nuance, or appealing to shared values—often amplify the very messages they oppose. This creates a tragic loop. The more sincerely one engages, the more one feeds the system that undermines sincerity. The hustler/sucker distinction thus maps directly onto platform-mediated political communication. Hustlers design messages for circulation while suckers engage for meaning.

The culmination is once more tragic as democratic agency requires trust in communication norms that no longer reliably hold. Citizens must act *as if* reasons matter even when they know that reasons are often secondary to identity and attention. To disengage entirely is to abandon democratic agency. To engage fully is to risk exploitation. No position of innocence here. The hustler/sucker architecture shifts responsibility away from individual voters and toward structural conditions. It challenges the narrative that democracy fails because citizens are stupid or misinformed. Instead, democracy falters because it demands participation under epistemic conditions that make exploitation rational. Hustlers succeed because democratic agency itself requires vulnerability.

5.5 Bullshit, Hustlers and Suckers

A direct line may be drawn between fake news, political communication and bullshit in Harry Frankfurt's sense of the word (Frankfurt 2005). Firstly, fake news may be defined as either downright lies or bullshit but it "simulates to be journalism and truth-seeking, while its goal is something entirely different" (Hendricks & Vestergaard 2019: 63). Now on a first reading Frankfurt's seminal essay is a very clever taxonomy of linguistic vice where liars acknowledge the truth/falsity distinction but deliberately distort it. Bullshitters, on the other hand, are indifferent to truth/falsity altogether. The distinction between liar and bullshitter when placed alongside the hustler/sucker framework takes on deeper significance. Bullshit emerges not simply as a defect of sincerity, but as a structural tool of exploitation under epistemic asymmetry.

To reach this conclusion, recap first the basic tenants of bullshit. According to Frankfurt, the bullshitter does not lie by asserting what they believe to be false. Rather, they speak without regard for whether what they say is true or false. Their primary concern is the perlocutionary effect of their speech—how it positions them, how it persuades, how it impresses.

This maps cleanly onto the figure of the hustler. The hustlers are not committed to falsehood but to *strategic effect*. They do not need their claims to be true, nor even believed in a robust sense. All they really need them to do is to circulate around, to provoke reaction, to incite indignation or to coordinate expectations. Truth is neither friend nor foe but rather irrelevant.

Frankfurt emphasizes that the bullshitter is often far more epistemologically dangerous than the liar. The liar still acknowledges the authority of truth by trying to conceal it. The bullshitter erodes that authority entirely. Considering the matter from the point of view of hustlers and suckers, this erosion is precisely what enables exploitation. When truth no longer functions as a shared constraint, those who continue to treat speech as truth-aimed become vulnerable.

So the sucker, in this reframing, is not the one who believes bullshit content. The sucker is the one who treats discourse *as if* it were governed by norms of sincerity, accuracy, and mutual understanding. This distinction is essential. Many victims of bullshit are not credulous. They may be skeptical, ironic, or even openly sardonic. What makes them suckers is not belief, but participation under the assumption that meaning still matters symmetrically. They respond, correct, argue, and engage *as if* reasons can settle things. Frankfurt notes that bullshit thrives where audiences care more about impression than accuracy. But the deeper tragedy lies in the fact that someone must continue to care about accuracy for communication to function at all. The sucker bears this cost. They sustain the background norm that speech is answerable to truth, even as others exploit that norm without honoring it. Thus, the sucker is not fooled by bullshit but are left exposed by their fidelity to discourse itself.

Frankfurt's analysis implicitly assumes an asymmetry between bullshitter and audience. The bullshitter knows they are bullshitting. The audience does not know whether the speaker is bullshitting, lying, or speaking sincerely. This asymmetry is not merely informational, again it is higher-order epistemic. This is exactly where the connection to non–common knowledge becomes explicit. Bullshit flourishes precisely because the speaker knows they are indifferent to truth, the listener does not know this, and crucially, the listener does not know whether others know this.

In other words, bullshit operates in environments where *sincerity is not common knowledge*, even if reluctance, suspicion or even skepticism is widespread. Many participants may privately suspect bullshit. Yet they cannot coordinate on treating it as such without risking social or strategic loss. This situation amounts to the exact epistemic structure of the hustler/sucker equilibrium. The hustler (bullshitter) exploits the fact that others must continue to act *as if* speech is truth-aimed in order to remain agents within the system. The sucker continues to play by discourse norms because abandoning them unilaterally would collapse their capacity to act meaningfully.

Frankfurt's notion of a "bull session"—an informal social context where participants speak freely without commitment to truth—offers a revealing case. In a bull session, statements are made for effect, bonding, or performance. Truth is explicitly bracketed. At first glance, the bull session seems harmless. Everyone knows what is going on.

But Frankfurt notes that bull sessions often involve *unequal awareness.* Some participants understand the game while others treat contributions as more sincere or informative than intended. This reveals the core asymmetry:

- The hustler knows the conversation is a bull session.
- The sucker does not know whether it is.
- Or even worse, knows but cannot act on that knowledge without social cost.

The faulty bull session thus exemplifies non–common knowledge. It is not that no one knows the discourse is unserious. The trouble is that this knowledge is unevenly distributed and not mutually acknowledged. The hustler exploits this by speaking without commitment while benefiting from others' interpretive labor. In contemporary public discourse, political communication, social media, finance, the bull session has become generalized. Entire arenas on social platforms operate as bull sessions without declaring themselves as such. Hustlers flourish as suckers continue to engage seriously.

Seen through the hustler/sucker lens, bullshit is a form of epistemic exploitation. It extracts value from others' commitment to meaning without reciprocating that commitment. In one end, the bullshitter benefits from attention, benefits from reaction, benefits from credibility spillover while refusing accountability to truth. At the other end, the sucker invests interpretive effort, risks reputational and cognitive cost, and sustains the communicative environment. Bullshit does not require belief—it requires *response*. Fact-checking, outrage, correction, parody—all feed the bullshitter's goals. The sucker's very attempt to restore truth amplifies bullshit's reach. This is why Frankfurt's analysis is so relevant to the tragedy of modern agency. Bullshit thrives not because people are stupid, but because agency requires engagement, and engagement under asymmetric sincerity is exploitable.

The final tragic insight emerges clearly. Modern agents face a dilemma. To disengage from bullshit is to risk silence, irrelevance, or withdrawal from public life. To engage is to risk being exploited by those indifferent to truth. There is no position of innocence. The sucker cannot simply stop caring about truth without abandoning agency. The hustler cannot restore sincerity without sacrificing advantage. Insight does not dissolve the structure only clarifies its cost. Frankfurt suggests that bullshit represents a deeper threat to truth than lying. The hustler/sucker framework shows why. Bullshit transforms truth from a shared constraint into a unilateral burden. Someone must still care, and that someone will pay—somebody must be a sucker. The bull session, once a marginal social form, has become dominant. Public discourse increasingly operates without shared commitment to truth, while still relying on some participants to behave as if such commitment exists. In a world saturated with bullshit—or in Frankfurt's words a world in which bullshit is such a "salient" feature—the tragedy is not that truth disappears. It is that truth becomes a liability borne unevenly by those who still care.

5.6 Marketing Bullshit

Marketing and advertising are often criticized for manipulating consumers through misleading framing like emphasizing gains over losses, hiding costs in fine print, anchoring prices, or presenting choices in ways that steer decisions without altering underlying value (Kahneman & Tversky 1984).

Behavioral economics explains these effects in terms of cognitive biases and heuristics (Thaler & Sunstein 2008). Consumers appear as predictable victims of framing; marketers appear as clever exploiters (Ariely 2008). While this account indeed captures important psychological mechanisms, there is additional structure. Misleading framing persists even when consumers are *aware* of being framed. People know about discounts that are not real discounts, "limited-time offers" that are permanent, subscription traps designed to discourage exit and other bullshit. Yet they still respond.

The hustler/sucker distinction explains why. Framing effects succeed not merely because people misperceive information, but because agency under market conditions requires acting under epistemic asymmetry, where refusing to engage is often costlier than risking being misled.

The sucker is the agent who continues to treat market communication as *meaningfully informative*, even when they suspect strategic distortion. Consider common framing practices like "Was $199, now $99", "Only 3 left in stock", "Free trial, cancel anytime" and "Recommended plan". Consumers frequently doubt the literal truth of these claims. Yet they still use them as *decision cues*, because ignoring them entirely would require unrealistic effort, time, or information. In markets saturated with options, heuristics are not cognitive failures; they are *functional necessities.* The sucker's vulnerability lies in the fact that *someone must interpret signals* for markets to function. The consumer supplies this interpretive labor, knowing it may be exploited, because opting out entirely would mean paralysis or exclusion.

The hustler in marketing is not simply a liar but more akin a bullshit artist. Modern advertising often relies on *choice architecture* (Kahneman & Tversky 1979, 1984, Betakova et al. 2025) structuring information so that certain interpretations become salient while others recede. Classic examples include:

- *Price partitioning* (low base price, high fees added later),
- *Decoy pricing* (introducing inferior options to make others seem attractive),
- *Loss framing* ("Don't miss out") instead of gain framing,
- *Default enrollment* in subscriptions or add-ons.

These strategies exploit the fact that consumers form expectations about others' behavior and market norms. A "popular choice" label works not because it proves quality, but because consumers believe others will choose it, making deviation feel risky. The hustler's advantage lies in *anticipating* how expectations will be formed, not in persuading consumers of specific truths.

Not surprisingly, misleading framing thrives under non–common knowledge. Consumers may individually recognize manipulative tactics, but they do not know whether others will act on this recognition. As long as enough others respond to frames, it remains rational to respond as well. Consider limited-time offers. Many consumers know that urgency is artificial. But if they believe others might buy now, waiting risks missing out. Thus, the frame remains effective even when disbelief is widespread. This mirrors the hustler/sucker equilibrium because as

- the hustler designs frames assuming consumers will act under uncertainty about others,

- the sucker responds because unilateral skepticism does not change market outcomes.

Awareness campaigns about manipulative advertising often fail for this reason. They create private insight without producing coordinated resistance.

Sales funnels illustrate the dynamic particularly clearly. Customers are led through small commitments—clicks, sign-ups, trials—that are individually low-cost and rational. So each step increases psychological and practical commitment. Examples include "Just enter your email", "No credit card required" and "Upgrade anytime". The framing minimizes perceived risk at each stage. By the time costs become salient, exit is effortful. Consumers often realize the structure midway but continue anyway, rationalizing further engagement to justify prior investment. Cognitive dissonance stabilizes suckerhood. Admitting exploitation would mean admitting wasted effort. Continuing allows the consumer to preserve a sense of agency and competence.

Responding to framing often involves *trading wits for convenience*—or for *vanity*. Personalized ads promise relevance, status, and identity affirmation. "This is for you" replaces "this is cheap." Consumers accept being tracked, nudged, and framed because these systems also offer ease, speed, and recognition. To reject framing entirely would require rejecting personalization, convenience, and social legibility. Thus, suckerhood is not simply imposed. It is *co-produced* by agents who value participation over distance. Hustlers simply design systems that monetize this trade.

5.7 Come Along for the Ride

A central implication of attention economics is that universal epistemic vigilance is impossible. If everyone treated all attention as suspect, discounted all visibility and refused all heuristic inference, then coordination would collapse, public discourse would freeze and collective action would become impossible and democracy die out. So once more, the attention economy structurally requires suckers. Not fools—but agents willing to trust signals, follow salience and participate without full higher-order knowledge. Hustlers exploit this very necessity.

The connection to financial bubbles now becomes clear. In markets price movements capture attention, attention attracts buyers and in turn buyers reinforce price movements. Crypto, meme stocks, and viral assets are attention-driven epistemic cascades. The same logic governs political misinformation, cultural virality, platform fame status economics (Hendricks & Hansen 2016) fake news, political communication, bullshit and marketing and advertising campaigns. Attention hustlers ride the cascade. Attention suckers mistake the cascade for collective belief. In essence, this is Aumann's agreeing-to-disagree problem, operationalized through attention.

Pairing up attention economics with the hustler/sucker distinction furnish a thorough explanation as to why the hustler/sucker distinction has become so salient in modern society. *In an attention economy, epistemic asymmetry is not accidental—it is profitable.* Hustlers are those who understand how attention shapes belief under scarcity. Suckers are those who must rely on attention cues to function socially. Social platforms institutionalize this asymmetry.

CHAPTER 6: Hustler/Sucker Platforms

The hustler/sucker distinction names something structurally important about modern social order. The way social interaction increasingly takes place under conditions of epistemic asymmetry, strategic opacity, and fragile trust already laid bare. By extension, hustlers and suckers are not personality types but roles produced by modern social conditions. They arise wherever individuals must act under uncertainty about others' beliefs, incentives, and rationality—conditions that are endemic to contemporary societies. The distinction captures a core tension of modernity as social life requires trust and norm-following, yet these same features generate persistent opportunities for exploitation. And social platforms are in on it.

V. F. Hendricks, *Sucker Nation*,
https://doi.org/10.1007/978-3-032-22788-1_6

6.1 Breakdown of Shared Frames

A sucker is someone who acts on beliefs about the social world that make them exploitable—typically beliefs about sincerity, cooperation, or shared norms. A hustler is someone who recognizes these beliefs as beliefs, rather than as facts about the world, and strategically exploits the gap between belief and reality. Neither role requires stupidity or brilliance. Both can be occupied by intelligent, educated, rational agents. Rehearsing from previous chapters what differs is what they know about what others know.

Modern societies are characterized by scale, anonymity, complexity and speed to "move fast and break things" (Taplin 2017, McNamee 2019). These features erode shared frames of interaction. In small-scale societies, roles, reputations, and norms are relatively transparent. Deviations are quickly detected and belief asymmetries are short-lived. In such contexts, hustlers and suckers are unstable roles.

By contrast, modern societies are by and large and impersonal, mediated by abstract systems like opaque markets, social platforms and bureaucratic constructions all of which saturated with signals whose meaning is ambiguous like prices, metrics, speech acts, narratives. The signal to noise ratio is overwhelming making it inherently difficult to know what others believe, who is sincere and who is strategic and whether apparent consensus is real, feigned or just simply performative. The result is a structural proliferation of epistemic asymmetries but now kept and controlled by social platforms. The hustler/sucker layout is unfortunately not the exception, but rather the default lens for interpreting and understanding social interaction.

Take something like trust. Trust is indispensable to modern society. Markets require trust in contracts, money, and institutions. Democracies require trust in procedures and information. Social life requires trust in language, promises and roles. At the same time, trust is strategically exploitable. Trust lowers cognitive and social transaction costs, but it also creates openings for those who do not reciprocate it. This produces a central paradox of modern life. The very norms that make large-scale cooperation possible also make exploitation possible.

Suckers are those who supply trust under the assumption that others do as well. Hustlers are those who consume trust without supplying it, while knowing that this asymmetry is not visible. Yet, modern societies cannot eliminate trust without collapsing nor may they eliminate hustlers without destroying the very opacity on which trust depends.

What distinguishes the hustler is not deception *per se*, but a special kind of *higher-order knowledge*. Hustlers know that social interaction is strategic even when it presents itself as non-strategic. They know that norms are locally binding but globally fragile. They understand that many people again act—*as if*—sincerity, fairness, or value are shared, even when they are not. In this sense, hustling is a form of second-order rationality in the sense of not just choosing well, but choosing well given others' mistaken models of the situation. This is the very reason why hustlers often appear cynical, ironic, or disenchanted. They see the game *behind* the game. Suckers, by contrast, are fully competent within the official script but blind to its strategic exploitation.

Why don't people simply stop being suckers? Because the alternative is worse. To refuse the sucker position entirely is to treat every interaction as potentially *adversarial*, every signal as strategic, every promise as empty, every word uttered or post encounted are parts and parcel of some hustle one way or the other. This is socially corrosive and practically impossible. It undermines coordination, cooperation, and meaning itself. Thus, being a sucker is often not a mistake, but a *necessary risk* of participation in social life. One can minimize vulnerability, but one cannot eliminate it without opting out. Modern discourse that glorifies "never being the sucker" often quietly presupposes that *others* will continue to play by norms. So universal hustling seems self-defeating.

6.2 The Industrialization of Hustling

In contemporary capitalism, hustling becomes institutionalized. Financial markets, advertising, marketing, influencer economies, and social platforms systematically produce conditions of epistemic non-transparency. Prices, likes, followers, and trends signal collective belief without revealing its composition. Individuals see enthusiasm but not skepticism, momentum but not exit intentions. This allows hustlers to operate at scale, while suckers mistake visibility for consensus. Financial bubbles, crypto markets, meme stocks and day trading were extreme but revealing cases. They show how large populations can be simultaneously rational, skeptical, and yet collectively exploitable—because belief asymmetries are not common knowledge.

The hustler/sucker distinction persists because it reflects a *structural tension* rather than a remediable flaw. Modern societies require trust but generate opacity. Opacity enables strategic exploitation. Exploitation reinforces suspicion but cannot eliminate trust. The result is a dynamic equilibrium in which hustlers and suckers continually reproduce one another.

Periods of scandal, crisis, or crash temporarily reduce sucker populations by making skepticism common knowledge. But over time, trust rebuilds, new narratives emerge, and the cycle resumes. How to get the cycle to resume itself again and again is the business of the social platforms relying on attention economy and digital capitalism.

6.3 Hustler/Sucker Machines

Attention economy and digital capitalism have not merely introduced new markets or technologies. It has likewise transformed the *epistemic conditions of social interaction.* Social platforms mediate how people see one another, how value is signaled, how success is measured and how collective belief is inferred (Taplin 2017). In doing so, they have made the hustler/sucker distinction not just common, but structurally central to contemporary life online and offline as the latter distinction slowly vanishes.

In platform society, everyone is warned not to be the sucker—yet the system depends on many people continuing to behave *as if* trust, authenticity, and shared understanding prevail. This is not a contradiction but a design feature. Platforms thrive by producing non–common knowledge sustaining large-scale hustler/sucker equilibria.

Digital capitalism differs from earlier forms of capitalism not primarily in ownership structures, but in how belief is organized and observed. Social platforms aggregate behavior (likes, clicks, shares, prices), display this behavior as signals of popularity or value, while obscuring the motivations, beliefs, and exit intentions behind it. This produces what might be called *epistemic compression*: Rich, heterogeneous belief states collapsed into simple metrics. Individuals see *that* something is popular, but not necessarily *why*, nor whether others believe the signal reflects genuine belief or some kind of strategic behavior. Not surprisingly, this is the ideal environment for hustler/sucker roles to emerge.

In platform society, hustlers are not merely individuals—they are, like before, positions within an architecture. Examples range from influencers monetizing attention while privately disbelieving their own brand narratives; founders and early investors promoting "community" while planning exit; growth hackers exploiting algorithmic visibility and even platforms themselves, extracting data to offer campaigns to interested advertisers, profile users, predict future behaviors or needs (Vestergaard 2019, Zuboff 2019) while framing participation as empowerment (Haugen 2021).

What unite these platform actors is higher-order knowledge about how signals are generated, how others interpret them while also knowing that this understanding is not common knowledge. Platform hustlers do not need to lie outright. They simply need to let others misinterpret visibility as value, engagement as belief, and participation as endorsement.

Conversely, platform suckers are not just dense users. They may very well be competent participants who act rationally within a distorted epistemic environment. They take metrics as social facts rather than strategic artifacts, treat visibility as consensus and often rather than not interpret participation as commitment. The error of the platform sucker is believing that others' behavior transparently reflects others' beliefs. This is precisely the higher-order belief error identified earlier. Platform users often know that manipulation exists—but do not know how widespread strategic behavior is, nor who knows this, or who knows that others know. As a result, they continue to supply attention, labor, data, and trust.

Influencer economies exemplify the industrialization of hustling. Influencers monetize authenticity, perform sincerity while strategically managing audience belief. Followers are encouraged to "trust" the influencer as a person, even though the relationship is structurally asymmetric and monetized. The sucker role is treating a market relation as a *para-social* one (Farivar et al. 2020, Balaban et al. 2022). At the same time social platforms intensify this by rewarding engagement regardless of sincerity, penalizing transparency when it reduces performance or engagement while making possible exit intentions largely invisible. Trust becomes a commodity—and exploitation becomes scalable.

Further to this end, social platform metrics create the illusion of collective agreement without producing common knowledge. A trending topic, viral post, or rising asset signal that many others are paying attention, but not whether they believe, mock, speculate, or plan to exit. This in turn allows users to believe that "others must really believe this," while others in turn believe the same about them. The result is a self-reinforcing epistemic loop—identical in structure to financial bubbles—now generalized across culture, politics, and identity.

Social platforms themselves occupy the hustler position. They know exactly how algorithms shape visibility, know that users may misinterpret signals and importantly know that this misinterpretation is not common knowledge. Yet they frame participation along lines of empowerment, community and democratization. Users are not deceived about terms of service or community standards. They are deceived about the *epistemic meaning of participation.* They believe they are engaging in social life, while platforms monetize this engagement as raw input. This is hustling without a hustler—structural exploitation embedded in design (Schüll 2012).

Platform discourse often urges users to "be savvy", "optimize", "think like a creator" and "build a personal brand" (Gandini 2020). This encourages universal hustling. But universal hustling is impossible as already made clear. Someone must supply *authentic* attention, someone must believe in the narrative and someone must behave as if norms apply. Thus platforms depend on a continual influx of suckers—new users, new creators, new believers—while rewarding those who learn to exploit the system. The result is a cultural atmosphere of cynicism without exit, some measure of awareness without coordination and finally suspicion without common knowledge.

In the end, what platform society systematically undermines is common knowledge. For all that has been said, users may individually know that metrics are manipulable or influencers monetize trust or that markets are speculative. What users lack is knowledge that others know and knowledge that others know that they know. Public space is considered *the* democratic arena for common knowledge to strive (Hendricks & Hansen 2016, Pinker 2025).

6.4 Public Space on Private Hands

In modern democracy, public spaces play a constant and fundamental democratic role in the provision of mutual recognition between citizens. In these public spaces, physical presence alone may lead to mutual recognition. The simple physical representation of the whole of society's citizens, in all their forms and appearances, establishes a public signal which inscribes itself in our collective consciousness. Society ladies with fanciful hats, the homeless on their cardboard boxes, children acting up, elderly with walking frames, sweaty schoolteachers with kids chasing pigeons, yuppies sipping cocktails, students with blue lit faces from Instagram's reflection glare and workers with their raw fists wrapped around hotdogs—they all have a right to a place in this consciousness. We are all represented through our collective consciousness and common background knowledge, as these form the basic premises for a democratic debate the framework of which is based on equality, freedom and mutual respect. The common value of public space is a fundamental value for democracy, whether we think of the City Hall Square in New York City or the steps of the Lincoln Memorial that provided the backdrop for "I Have a Dream".

When public space serves as a fundamental democratic pillar, it is a central premise that no one may claim special interest in or capitalize from it, as it would no longer be equal for all nor provide equal access. The past 15–20 years have seen a paradigm shift in our understanding and ownership of the public space. In conjunction with public space increasingly becoming digital infrastructure, our temporal physical presence in the same physical space has become less important.

It is not democratic nations or supranational organizations that have facilitated this massive transformation of public space, but private enterprise and primarily Big Tech companies like Alphabet, Amazon, Microsoft, Apple, X and not least Meta/Facebook running the world's largest town hall. These tech companies however do not answer to founding democratic principles, but instead to their advertisers and stockholders. Leaders of such companies are on record for saying things like safeguarding democracy is "above our pay grade" (Zuckerberg 2019) or "We will coup whoever we want! Deal with it" (Musk 2023) and "I no longer believe that freedom and democracy are compatible" (Thiel 2025). Sounds suspiciously like a one-way ticket to "Dark Enlightenment".

Dark Enlightenment is a far-right intellectual movement, also called the Neo-Reactionary movement (abbreviated to NRx). The movement seeks to replace democratic governance with a techno-authoritarian or techno-feudal, corporate-style state run by a powerful executive (Simon 2025). Dismissing egalitarian ideals and liberal democracy as fundamentally flawed Dark Enlightenment instead advocates hierarchies and centralized control. Its ideas, largely circulated online and influential among some tech and political elites, critique modern democratic institutions and promote a vision of governance where the state operates more like a business with a "CEO" at the helm, reflecting deep skepticism about the Enlightenment's legacy of liberty and progress.

Back to the microscopic perspective on public space on private hands. It may very well be that all citizens have gained a voice, in so far that virtually anyone can have a profile on a social platform. What is of interest, however, is not that we've all gained a voice but whether our voice is heard. Some people and some things are heard and seen more than others in the information market ruled by the attention economy; hence some privileged parties have a greater possibility of becoming part of our collective consciousness and decide what is to be common knowledge and what is non-common knowledge in a public space on private hands. That's power—lots of power to wield. Because the understanding of public space's central function as a foundation for common knowledge as well as the inherent possibilities for creating and controlling information phenomena, provides not only the basic concepts for understanding a society's fundamental fellow-feeling, but also for the political power ruling on this very basis. As scientist and philosopher Michael Polanyi has very pointedly described it in epistemic terms: "[...] If in a group of men each believes that all others will obey the commands of a person claiming to be their common superior, all will obey this person as their superior [...][A]ll are forced to obey by the mere supposition of the others' continued obedience." (Polyani 1958: 224)

Much may be said about democracies but they've seldom, if ever, been decidedly pecuniary profitable. Democracy is the societal model that we in the Western world have thought of as the pre-eminent way of securing both individual and collective interests while keeping checks and balances on the powers that be. With the dominance of

Big Tech it's worth asking whether public space should be left to private enterprise — or if it in reverse is downright detrimental to the individual as well the collective? Had Kant, Rousseau or Montesquieu, who contributed to carving out the philosophical path for modern democracy back in the Enlightenment, been consulted on whether the infrastructure of public space should be on private hands, their answer would have been a firm and resounding—no thanks! There can't be special interest, and no one should be able to capitalize on the democratic infrastructure if the marketplace of ideas is to be free and equally accessible to all. But only a sucker acts as if the marketplace of ideas is truly free in a public space on private hands ruled by digital capitalism or rather technofeudal lords.

6.5 Capitalist Exchange vs. Feudal Rent

Digital capitalism and platform economy is increasingly described not as competitive or entrepreneurial, but as feudal. The term *technofeudalism* captures a growing intuition. That economic life is no longer organized primarily around markets and contracts between equals, but around platforms that function as private infrastructures, extracting rents from those who depend on them (Varoufakis 2024). What distinguishes technofeudalism from earlier forms of domination is not coercion in the classical sense, but *dependency without exit.* One does not work for the lord; one lives on the lord's land—just deal with it is the only option at hand.

The metaphor of *paying rent on one's own face* names precisely the subjective and epistemic dimension of this condition. In technofeudal societies, individuals do not merely pay rent in money. They pay rent in visibility, attention, identity, status and self-disclosure. To appear as a social agent at all—to speak, to be heard, to count—one must continually expose oneself to social platforms and systems that own the terrain of appearance. The face becomes leased property.

In classical capitalism, exploitation occurs through exchange. Workers sell labor and capitalists extract surplus value. The relation is asymmetrical, but conceptually transparent. In technofeudalism, by contrast, the central economic relation is rent, not exchange. Platforms do not primarily sell products or even services. They control access while collecting attention. They practically own the digital land on which social, economic, and political life now takes place.

To participate, one must pay. But the payment is not primarily monetary. It is continuous exposure in terms of posting, reacting, authenticating, signaling and remaining legible. This is not a transaction one can complete and exit. It is an ongoing condition of presence. To stop paying rent is by and large to disappear.

The "face" in the metaphor above is not merely one's image. It is one's socially recognizable self including beliefs, emotions, values, identity and reputation. In predigital societies, one could appear selectively, in bounded contexts, before limited audiences. Today, appearance is mediated almost entirely by privately owned infrastructures whose incentives are opaque and extractive. To have a face in public is to submit it to metrics, allow it to be parsed by algorithms, let it be recombined, amplified, or buried (Galloway 2017).

The tragedy is that the face still feels personal. One experiences exposure as self-expression, participation or authenticity. Yet from the platform's perspective, the face is not a person—it is a *resource stream* for which *all extracted data is credit information* (Vestergaard 2019, Hendricks 2023) This is essentially where techno-feudalism departs decisively from capitalism. The subject is not only exploited as a worker or consumer, but as a tenant of their own appearance.

6.6 Hustlers and Suckers in Technofeudalism

Technofeudalism produces its own social roles, and the hustler/sucker distinction maps onto them precisely. The sucker is the tenant who pays rent sincerely. As before, they believe that visibility is recognition, authenticity is empowerment and that participation amounts to agency. They trade strategic distance for the promise of being seen.

Their rent payments take the form of attention, outrage, self-disclosure, and moral labor. They are just complying with the conditions of tenancy and that's all there is to it. *Suck on it.*

The hustler, by contrast, understands the feudal structure. Hustlers, like before, minimize their own exposure, monetize the exposure of others, perform authenticity without surrendering control, treat the face as an asset rather than a self.

Influencers, outrage entrepreneurs, rage baiters, early adopters, and platforms themselves occupy this position. They do not abolish rent. They collect it kicking much of it back to what Varoufakis refers to as the *cloudalists* or *technofeudal lords* owning and administering the information borne infrastructure at large. Recall that hustlers and suckers are not moral opposites. They are complementary roles within the same system. The hustler requires the sucker's sincere rent payments. The sucker depends on the system that the hustler exploits.

The mapping to the hustler/sucker distinction goes even further because what stabilizes technofeudalism is not ignorance, but the lack of common knowledge. We very well may know that platforms extract value from visibility, that authenticity is monetized, that outrage is profitable and that data is everything when it comes to extracting credit information about users, companies, institutions and governments. What we do not know is whether others are ready to stop paying rent—whether withdrawal would be collective or merely self-erasing. This mirrors classical feudalism, where peasants often understood their exploitation but lacked coordination and exit. In technofeudalism, exit is epistemic rather than physical. One can leave a social platform, but not the infrastructure of appearance itself, which is fragmented across multiple private domains. Thus, people continue to pay rent, not because they are deceived, but because refusal appears futile.

In technofeudalism, authenticity becomes not a virtue but an obligation. Social platforms demand continuous self-revelation in terms of being real, being yourself and speaking your mind. Authenticity thus functions like a *feudal tithe*. The tenant must regularly present their face to the lord—not as labor, but as *proof of presence*. The more authentic the exposure, the higher the rent. The hustler understands this obligation and exploits it. The sucker internalizes it as ethical necessity.

6.7 Agency under Technofeudalism

The deepest tragedy of technofeudalism is that agency itself becomes conditional on rent payment. To act, one must appear. To appear, one must submit. To submit, one must accept extraction. No innocent position as full participation entails exploitation while full withdrawal entails disappearance yet partial participation entails constant calculation. This is exactly the reson why technofeudalism produces neither rebellion nor consent, but resigned compliance. People know the system is extractive, but cannot see how to coordinate escape without losing their face altogether.

Technofeudalism cannot be dismantled by individual

refusal alone, just as feudalism could not be abolished by isolated peasants. What can be done, first, is to name the structure. The metaphor of paying rent on one's own face clarifies what is otherwise experienced as vague discomfort, burnout, or cynicism. It reveals that modern subjects are not merely overworked or misinformed, but *ontologically indebted*—required to continually justify their presence by making themselves visible and exploitable. Understanding this does not free us. But it shifts the moral burden. It shows that vulnerability is not personal failure, that cynicism is not wisdom, and that hustling is not liberation. All are responses to a system that has replaced exchange with rent, participation with tenancy, and appearance with extraction. If technofeudalism has a defining feature, it is this: We no longer pay to live on the land—we pay to appear as ourselves. And the rent is always due. That's the new way of wits, vanity even authenticity.

CHAPTER 7: Wits, Vanity and Authenticity

At first glance, calling suckers *vain* sounds like moral insult. Suckers are just shallow, narcissistic, ego-driven. The more accurate claim is that suckers are vulnerable because they overestimate the epistemic significance of their own visibility, participation, or moral stance. Vanity here is not self-love, but misplaced reflexivity—a systematic error about how one's own actions are perceived, valued, and interpreted in a strategic, attention-mediated yet attention scarce environment.

7.1 Vanity in the History of Philosophy

In the history of philosophy, vanity back in the day appeared as a moral failing in terms of excessive concern with appearance, reputation, or esteem. In classical and early Christian thought, vanity is condemned not because it is socially harmful, but because it represents a misalign-

V. F. Hendricks, *Sucker Nation*,
https://doi.org/10.1007/978-3-032-22788-1_7

ment of the soul—an orientation toward externals rather than truth or virtue.

Plato does not use the specific word "vanity", but he addresses the concept through related Greek terms like *philotimes* (meaning love of honor) and *alazoneia* (which is boasting or arrogance). A concern for reputation (*doxa* – in the sense of how a person appears or is judged by the public) is opposed to concern for truth. "The "timocratic" man in *The Republic* (Book VIII) is ruled by a desire for honor and standing rather than reason. In the ideal state, rule by reason is preferred over rule by honor or vanity (Plato 2007). *The Sophist*, who flatters audiences and seeks admiration, embodies vanity as epistemic corruption being more concerned with reputation than with truth (Plato 2015). Aristotle is more tempered and moderate, recognizing in the *Nicomachean Ethics* a legitimate concern for honor, but he too treats vanity (*kenodoxia*, meaning vainglory, empty conceit or groundless pride) as excess or a failure of measure in self-regard (Aristotle 2009).

Christian theology radicalizes this critique. Augustine in *Confessions* (1991) and later Aquinas in *Summa Theologiae* (1981) frame vanity as a spiritual disorder. The desire to be seen rather than to be good. Vanity distracts the self from God and traps it in worldly comparison. At this stage, vanity is an individual moral vice, correctable through humility and proper orientation of desire.

What is largely absent from these early accounts, save perhaps from Plato, is a sense in which vanity might be *structurally induced*—that social conditions could require individuals to attend to appearance in order to function as agents at all. The early modern period begins to shift the meaning of vanity from a purely moral failing to a socially mediated phenomenon. With the rise of commercial society, courts, and public life, reputation becomes increasingly consequential. One cannot simply withdraw from appearances—one must manage them. Thinkers such as Hobbes and Mandeville mark a turning point. In *Leviathan*, Hobbes recognizes *vainglory* as a basic human motive tied to competition and insecurity (Hobbes 1996). Mandeville, in the *The Fable of the Bees* goes further, arguing provocatively that vanity and pride are not merely unavoidable, but socially productive. They drive commerce, innovation, and social order (Mandeville 1989). Here vanity begins to lose its purely negative status. It becomes more instrumental as resource mobilized by social systems. Recognition, esteem, and visibility are no longer distractions from agency but conditions of participation.

This is a major step toward the hustler/sucker body-work. Once appearance becomes structurally relevant, those who continue to treat vanity as a moral failing are increasingly vulnerable to those who exploit it strategically. The Enlightenment deepens this transformation by tying vanity to *social comparison* and recognition. Rousseau is pivotal. In the *Discourse on Inequality* (1997) he distinguishes between *amour de soi* (healthy self-love) and *amour-propre* (comparative self-regard). Vanity, in the Rousseauian sense, arises when self-worth depends on the gaze of others. *Amour-propre* is not optional. It emerges inevitably in social life. Once individuals depend on recognition, they become vulnerable to misrecognition and manipulation. Rousseau does not yet describe hustlers and suckers, but he identifies the psychological terrain on which they will later appear. Kant attempts to contain this vulnerability by grounding dignity in moral autonomy rather than recognition. Yet Kant's moral subject is increasingly fragile in a world where social and economic participation depends on visibility, reputation, and performance. Moral dignity does not eliminate vanity's practical necessity (Kant 1996).

Nietzsche marks a significant rupture. He rejects the moral condemnation of vanity altogether, reinterpreting it as a symptom of deeper power dynamics. What earlier philosophers called vanity, Nietzsche reframes as the struggle for recognition and dominance (Nietzsche 1994). Yet Nietzsche also exposes the cost of this struggle. The modern individual is trapped between cynicism and self-display, irony and performance. Authenticity itself becomes performative. Vanity no longer names a simple vice but rather a *condition of social existence*. This prepares the ground for the hustler/sucker distinction. Once all values are seen as perspectival and strategic, the question becomes not whether vanity is bad, but *who pays its cost.*

7.2 Vanity Fair

The history of the philosophy of vanity charts a slow but determining shift. From moral vice, to social motive, to structural condition. What began as a personal failing becomes a systemic vulnerability. The hustler/sucker distinction names the final stage of this history. It shows how vanity—once condemned, later tolerated, now required—becomes the site where modern agency fractures. The one to connect the dots between vanity and attention economics is George Franck through his idea of the *vanity fair* (Franck 2020).

For Franck, modern society is organized around the production, accumulation, and distribution of attention as a scarce resource. Attention functions analogously to capital. It can be invested, accumulated, transferred, and converted into power, money, and influence (Franck 2018). In this framework, the vanity fair is no longer a place people occasionally visit. It is the permanent market structure of modern social life.

Franck's central insight is that attention precedes money. Before anything can be valued, sold, or legitimized, it must be noticed. Attention therefore becomes the most fundamental currency of modern societies. Importantly, attention is not evenly distributed, nor is it freely chosen. It is organized through media systems, platforms, and reputational hierarchies. In this economy, self-presentation becomes labor. Individuals must invest in visibility to maintain social existence. To be unseen is not merely to be ignored, but to be excluded from agency itself. This marks the end of innocence. One can no longer simply act—one must appear acting.

Vanity is redefined. It is no longer excessive concern

with appearance. It is the practical orientation toward being seen, without which action loses efficacy. Franck's vanity fair is therefore not populated by narcissists, but by agents compelled to manage their visibility in order to function at all. In vanity fair, the hustler is the agent who understands the conversion mechanisms of attention. Hustlers know that attention is not recognition, but extractable surplus. They treat visibility strategically, minimizing their own exposure while maximizing the exposure of others. Influencers, outrage entrepreneurs, platform architects, and symbolic elites occupy this position. They do not need to persuade others of truths or falsehoods. They need only to capture and redirect attention flows. Meaning is secondary—circulation is primary.

The hustler is the rational actor of attention economics. They understand that vanity is a resource others must supply. Hustling consists in positioning oneself as the beneficiary of others' attention investments. The sucker, by contrast, is not ignorant of the attention economy. Often they articulate its dynamics with clarity. What distinguishes them is that they continue to act *as if* attention were still recognition—*as if* being seen were being acknowledged rather than harvested.

Suckers invest in authenticity, visibility, participation, and exposure under the belief that these confer dignity, voice, or agency. In Franck's terms, they supply attention capital without controlling its conversion. Their vanity is not excessive but structurally induced. This is where the idea of trading wits for vanity re-enters. Fully acknowledging that one's visibility is being monetized would undermine the psychological conditions of action. To remain an agent, the sucker accepts the partial illusion that their appearance matters on its own terms.

Franck's analysis implicitly relies on an epistemic asymmetry. Many participants know that attention is being extracted. What they do not know is whether others are ready to stop paying. Lack of common knowledge again. Individuals may resent the attention economy, mock it, or critique it—yet continue to participate because withdrawal appears futile or self-erasing. Irony replaces exit. Awareness replaces coordination. The vanity fair thus persists because knowledge does not synchronize. The sucker continues to supply attention because unilateral refusal threatens invisibility. The hustler continues to extract because supply remains stable. That really makes vanity a bitc*.

7.3 Vanity is a Bitc*

In platform-mediated society vanity is treating one's own visibility as epistemically meaningful in a system that treats visibility instrumentally. This is an epistemic mismatch rather than a moral failing. The hustler understands something the sucker does not. Namely that visibility is not the same belief, expression is not identical to commitment and that participation is not tantamount to endorsement. Being a student well-versed in attention economy, the hustler also knows that attention is harvested, not respected.

Knows that every expression is monetized on social platforms, not necessarily interpreted sincerely and that moral or personal signaling is the raw material of the business. The sucker, by contrast, implicitly believes that "If I show myself—my values, my outrage, my enthusiasm—it *counts*." That belief is more often than not false in an attention economy. So the suckers' vanity consists in overestimating their epistemic centrality. Suckers assume that their reaction is noticed as a reason, that their participation somehow contributes to collective understanding and that their stance affects the epistemic environment positively. The problem is that in platform society, individual acts are aggregated, not interpreted, signals are stripped of intent while meaning is replaced by metrics. Thus the sucker's vanity is not narcissism, but a mistaken model of how meaning is processed.

For instance outrage works because it appeals to a specific vanity like "My moral reaction matters." Users share, comment, denounce, correct—not because they necessarily believe the content, but because they believe their *response* is epistemically valuable to themselves and others. But hustlers already know that outrage amplifies regardless of intent, even corrections feed visibility and that moral performance is attention fuel. So while the sucker's vanity trap is believing that moral expression resists manipulation, the hustler knows it accelerates it.

Vanity also explains why suckers fail to recognize higher-order belief structure. The reason is again much the same as before. Suckers assume others interpret their actions sincerely, they believe that others are able to distinguish belief from strategy and very importantly that others are "like them." This is a vanity of *epistemic projection* in the sense that "Others see the world as I do, and see me as I intend." The hustler does not share this illusion. They know that others are opaque whatever they are. On top of that interpretation is typically shallow and the social platforms and systems really do not care about intent. This asymmetry sustains the hustler/sucker equilibrium.

Vanity in this sense is not necessarily irrational. Social life typically requires acting as if one's words matter, acting as if sincerity is legible and acting as if participation is meaningful. Without these assumptions communication may very well collapse entailing that coordination would fail all the while agency would dissolve into cynicism. With the assumptions in place vanity becomes cost of being a social agent. Suckers are not foolish for being vain they are simply exposed because they must be.

Hustlers are not free of vanity. They simply locate it elsewhere. Their vanity lies in seeing themselves as above the game believing they understand others while remaining unseen and taking pride in strategic invisibility. But this vanity is *aligned* with the social platform or system, not against it. And it pays off: *The suckers believe they are seen. The hustlers know they are counted.*

7.4 Trading Wits for Vanity

The conquering thought of *Sucker Nation* is that modern society does not simply divide people into the clever and the foolish. Rather, it organizes social life around a subtler trade-off *between wits and vanity.* To navigate strategically, one must maintain distance, opacity, and restraint. To be visible, recognized, and affirmed, one must expose oneself, signal sincerity, and participate. The hustler/sucker distinction marks the fault line where this trade-off becomes explicit.

To be a sucker is to have exchanged strategic acuity for the promise of recognition—to have accepted vanity as compensation for vulnerability. To be a hustler is to refuse this exchange, or to profit from others making it.

Vanity is now *needing to be legible*—to have one's identity, values, and emotions publicly acknowledged. Vanity today is epistemic before it is moral consisting of a complex of beliefs including that my expression will be interpreted as intended, that my sincerity will be recognized as sincere and that my participation participation will "count" in some meaningful way. This belief set is not absurd. But under conditions of digital mediation and attention economics, it may become a liability.

To "trade wits for vanity" means accepting a bargain offered implicitly by platform society: Expose your beliefs, emotions, and identity, and you will be seen. Remain strategically cautious, and you will be ignored. The sucker accepts this bargain and while doing so gives up strategic ambiguity, epistemic distance and the option of silence. In return, suckers receive visibility, a feeling of participation and relevance, even moral affirmation. The hustler, by contrast, rejects the bargain—or accepts it only performatively. Hustlers maintain opacity while harvesting the attention of those who do not.

This exchange may not be as irrational as it seems on first reading. Social platforms punish invisibility (Srnicek 2016). Political relevance, professional success, cultural legitimacy, and even personal identity are increasingly mediated by visibility metrics. To opt out entirely is to risk social irrelevance. Thus, many users rationally choose vanity over wits. They speak when silence would be safer. They react when restraint would be wiser. They signal when ambiguity would protect them. And they produce opinions even if they are uninformed. The tragedy is that the cost of strategic intelligence has become social disappearance.

Hustlers succeed because they exactly understand the price of vanity. They know all too well that being seen is not the same as being respected nor that being loud is not the same as being influential and as it were being sincere is not the same as being safe. Hustlers in turn treat visibility instrumentally. They may *perform* authenticity, outrage, or vulnerability, but they do so without collapsing identity into expression. Their wits lie precisely in withholding themselves while encouraging others to reveal. This is why hustlers often appear cynical or ironic—irony is a form of distance, and distance is protection.

One reason suckerhood persists is because vanity is moralized. Participation is framed as courage—silence as complicity. Authentic expression is framed as virtue while restraint as inauthenticity. In such a moral economy, refusing exposure feels like a moral failure. The sucker does not merely expose themselves for attention; they do so to maintain moral self-respect. They believe something to the effect that "If I do not speak, I am failing." This moral pressure completes the trade. Wits are surrendered not only for visibility, but for moral reassurance. Hustlers exploit this pressure ruthlessly. It's bullshit anyway.

The pattern recurs across domains. In politics, citizens trade strategic patience for expressive outrage, believing their stance matters, while political hustlers convert outrage into attention and power. In markets, retail investors trade cautious skepticism for the excitement of participation, while insiders profit from their visibility-driven enthusiasm. In online culture, content creators trade long-term autonomy for algorithmic recognition, while social platforms monetize their exposure. In each case, suckerhood is not ignorance but overinvestment in being seen.

And why not? Social platforms sell vanity as empowerment anyway. "Speak your truth". "Build your brand". "Own your voice". Yet empowerment without control over interpretation is illusory. To trade wits for vanity is to *mistake expression for agency*. The hustler knows that agency lies elsewhere—in timing, in withholding, in understanding how others misinterpret visibility.

Refusing the trade entirely is costly. It risks invisibility, marginalization perhaps even loneliness. But refusing it selectively is possible. This essentially means choosing when visibility is worth the exposure, preserving silence where nothing is gained by speech and finally separating self-worth from public recognition. Such refusal is not anti-social—it is *anti-extractive*.

The hustler/sucker distinction ultimately reveals the tension between being seen and staying smart. To be fully visible is to be legible, predictable, and exploitable. To be fully strategic is to risk obscurity. Suckers are exposed because they have accepted a social bargain that rewards vanity and punishes restraint. Hustlers succeed because they refuse to confuse recognition with power. To stop trading wits for vanity is to reclaim the right to opacity, to silence, and to selective engagement. In a world that constantly demands exposure, the most subversive form of intelligence may be knowing when not to be seen. And that, perhaps, is the most important lesson of the hustler/sucker distinction.

7.5 Cognitive Dissonance

Cognitive dissonance theory, originating with psychologist, Leon Festinger, describes the discomfort experienced when actions and beliefs conflict (Festinger 1957). Individuals are motivated to reduce this discomfort, often by adjusting beliefs rather than behavior. In the hustler/sucker context—as alluded to on multiple occasions already—the dissonance is acute because the agents know exposure is risky and exploitable, continues to expose themselves and therefore needs some story that preserves self-respect. Vanity provides this story. The cognitive dissonance is resolved not by changing behavior, but by reframing vulnerability as virtue. So the sucker does not deny exploitation outright. They deny its *significance* by embedding their participation in a narrative of personal or moral worth.

"Trading wits for vanity" names this exact exchange. Wits represent strategic clarity. The capacity to see how one's actions are embedded in systems of extraction, manipulation, and asymmetrical payoff. Vanity represents the need to sustain a coherent, dignified self-image within those systems. To maintain full strategic clarity while continuing to participate would require accepting that one's actions are often instrumentalized, one's sincerity may not matter and one's visibility may benefit others more than oneself. This acceptance is corrosive to agency understood as self-authorship. It risks producing cynicism, paralysis, or alienation.

Vanity offers the relief. It allows the agent to believe that *my* participation is different, *my* voice matters and *my* authenticity protects me. In accepting this belief, the agent sacrifices some strategic clarity—but gains psychological coherence and social presence.

Hustlers experience less cognitive dissonance because they do not need to reconcile exposure with self-image. They preserve distance. Their self-concept does not depend on being recognized sincerely. They accept instrumentalization as the nature of the game. This does not make them happier or more free. It makes them *less internally conflicted.* Their dissonance is displaced outward, onto those who continue to act sincerely. Very importantly, hustlers often rely on the vanity of others. They benefit from environments in which exposure is moralized and authenticity celebrated. The sucker's dissonance management becomes the hustler's profit condition.

One reason suckerhood persists is that exiting the role just creates *new* dissonance. Withdrawal, silence, or strategic restraint often conflict with deeply internalized norms of participation, courage, and engagement. To stop participating is not experienced as neutral. It is experienced as either cowardice, disengagement, elitism or complicity. These moral framings intensify dissonance for those who might otherwise withdraw. Continuing to participate, even at a cost, allows the agent to preserve a positive self-image. Here again, vanity stabilizes the equilibrium.

The tragic insight is that *there is no dissonance-free position.* Full clarity without vanity leads to alienation. Full vanity without clarity leads to exploitation. Partial clarity and partial vanity lead to ongoing tension. Modern agency is therefore lived under *managed dissonance.* One must accept some self-deception to remain an agent at all. Innocence cannot be restored, pure strategic wisdom is corrosive and total authenticity is exploitable. So authenticity becomes a vanity trap.

7.6 Authenticity as a Vanity Trap

Few ideals are celebrated more loudly in platform society than authenticity (Haimson et al. 2021, Taylor 2022). "Be yourself". "Speak your truth". "Show what apparently is the real you". Across social platforms, branding, politics and culture, authenticity is presented as an ethical corrective to manipulation, fakery, and cynicism. Yet the same environments that valorize authenticity are also those in which it is most systematically exploited.

On social platforms, authenticity functions as a *vanity trap*. A norm that invites individuals to expose sincerity, identity, and moral stance under the illusion that doing so resists manipulation, while in fact making them maximally legible to systems and actors that monetize attention. Authenticity does not protect against hustling. It enables it. Historically, authenticity emerged as a moral ideal opposed to hypocrisy and alienation—most clearly in Rousseau (Rousseau 1755/1997), but also later in existentialism (Sartre 2003). To be authentic was to resist social falsification, to act in accordance with one's true self rather than external pressures.

Platform society inverts this reasoning. Authenticity is no longer resistance to social evaluation but the raw material for it. What was once inward-directed (fidelity to oneself) becomes outward-facing (visibility of the self). The authentic self must now be performed, circulated, and recognized. This shift turns authenticity into a strategic resource—and a vulnerability.

Among hustlers and suckers, authenticity performs a specific epistemic function by collapsing the distance between belief, expression, and identity. When agents present themselves as authentic, they reveal everything from preferences over values and emotional triggers to social alignments. This information is invaluable—not primarily to other individuals, but to social platforms and strategic actors alike. The authentic subject believes: "If I am sincere, I cannot be manipulated." The hustler knows: "Sincerity makes you predictable."

The vanity embedded in authenticity is a notion of self-significance in the sense of believing that one's actually sincerity matters. Believing that self-expression will be interpreted as intended and also—maybe to a less extent—that that one's participation resists commodification. This is an epistemic vanity: "My inner truth cannot be reduced to a signal." In platform society, this belief is blatantly false. Social platforms do not interpret authenticity as much as they *process* it. Authentic expression is by and large stripped of context, aggregated, and converted into engagement metrics. Meaning is irrelevant; visibility is everything. Hereby authenticity becomes a trap. The more sincerely one expresses oneself, the more efficiently one is exploited.

Hustlers understand authenticity very differently. They basically treat authenticity as a *style*, not a state, a *signal*, not a commitment and in the end a *resource*, not a value. Influencers, political actors, outrage entrepreneurs and rage baiters perform authenticity strategically in terms of curated vulnerability, scripted spontaneity and managed sincerity. They know that audiences reward authenticity *as appearance*, not as truth (Hendricks & Hansen 2016). They appear authentic while remaining strategically opaque. This may explain why hustlers often seem paradoxically more "real" than genuinely sincere participants.

Suckers, by contrast, take authenticity at face value. They routinely express outrage sincerely, share genuine experiences and signal identity more or less honestly. Why? Because such moral and emotional labor may actually improves discourse, be resistant to manipulation while building community. But in an attention economy, this labor is unpaid input. It fuels algorithms, amplifies narratives, provides data for targeting and credit information for profiling and trading futures of human behavior. The sucker's authenticity does not disrupt the system; it sustains it. Thus authenticity becomes a one-sided commitment binding for the sincere, optional for the strategic.

This dynamic also explains why authenticity is central to fake news ecosystems. Fake news relies on people reacting authentically in terms of outrage, fear and other activity mobilizing sentiments (Berger 2013) issuing moral condemnation or defense of identity. The sucker believes that "My authentic reaction fights misinformation." Yet the hustler knows that "Any reaction feeds the attention economy." So again, authenticity accelerates epistemic pollution by making emotional engagement morally obligatory yet epistemically naïve.

Calls for authenticity are often framed as antidotes to manipulation: Be real, be sincere, resist spin. In platform society, this advice is not merely ineffective—it is counterproductive. The reason is that authenticity only increases exposure while reducing strategic ambiguity and simultaneously lowers defensive distance. What protects against hustling is epistemic caution. As already noted this amounts to the ability to separate identity from expression, belief from visibility and participation from endorsement.

The deepest problem with authenticity is a recurring one. It presupposes a world in which others interpret sincerity sincerely, that social platforms respect intent and that meaning is not systematically distorted. This presupposition fails under conditions of non–common knowledge. One may know that authenticity is exploited—but not know whether others know, or know that one knows.

Thus individuals remain trapped. They become afraid to withdraw authenticity (lest they lose voice). Forget about coordinating restraint. They remain vulnerable to strategic exploitation. Authenticity persists as a norm because its costs are individual, while its coordination requires collective epistemic alignment.

The challenge for modern society is to decouple authenticity from compulsory visibility—to recover spaces where sincerity is not immediately rendered legible, countable, and monetizable. Until then, authenticity will remain what it increasingly has become. Not a virtue that frees us, but rather a vanity trap that binds us ever more tightly to the systems exploiting our natural need to be seen. That's the existential predicament of our time and the true tragedy of modern agency.

CHAPTER 8: A Tragedy of Modern Agency

The distinction between hustlers and suckers may appear, at first glance, too colloquial to deserve philosophical treatment. It seems to belong to the vocabulary of profanity rather than the lexicon of the academy, street wisdom rather than theory, to informal judgment rather than conceptual analysis. And yet, once examined closely, the distinction reveals itself as one of the most persistent—and most revealing—ways in which we have tried to think about agency under conditions of asymmetry. The hustler/sucker distinction has indeed philosophical prehistory.

8.1 A Short History of Sucker Philosophy

Across the history of philosophy it has been routinely asked: How should one act in a world where others may exploit

V. F. Hendricks, *Sucker Nation*,
https://doi.org/10.1007/978-3-032-22788-1_8

trust, manipulate appearances, or free-ride on cooperation? At different moments, this concern is framed morally (as virtue versus vice), politically (as power versus vulnerability), or epistemically (as knowledge versus ignorance). What changes is not the problem, but rather the conditions under which it appears.

In ancient Greek philosophy, the hustler/sucker distinction appears, to the extent that it does, primarily as a *moral problem* (Sophocles 2003, 2008). The exploiter is unjust or vicious and the exploited is foolish or insufficiently virtuous. Now this framing presupposes that suckerhood is somehow *correctable*. In Plato's dialogues, figures like Thrasymachus (Plato 2007) or the Sophists often resemble proto-hustlers (Plato 2015). They are rhetorically skilled, strategically minded, and willing to exploit appearances for advantage. Against them stands Socratic philosophy, which insists that injustice is ultimately self-defeating. To exploit others is to harm one's soul. The sucker's vulnerability is essentially explained not by structural conditions, but by lack of philosophical insight.

Aristotle develops this view more systematically (Aristotle 2009). For him, practical wisdom (*phronesis*) allows agents to navigate social life without falling prey to manipulation. The vicious person exploits while the foolish person is exploited. The virtuous person avoids both extremes by acting appropriately within shared norms. Trust, here, is not tragic—it is calibrated by character.

What is absent from such philosophical thinking is any sense that virtue itself might be systematically exploitable.

The ancient world presupposes relatively bounded communities, shared norms, and face-to-face accountability. Under these conditions, hustlers can be exposed, suckers educated, and justice restored. The tragic dimension has yet to emerge.

The early modern period marks the quintessential shift. With the breakdown of medieval moral unity and the rise of large-scale political conflict, the hustler/sucker distinction becomes *political and strategic.* Machiavelli is the seminal figure here. In *The Prince*, he explicitly rejects the assumption that virtue protects against exploitation (Machiavelli 1981). On the contrary, those who act morally in a world of strategic actors are likely to be destroyed. The sucker is not imprudent but *unrealistic.* The hustler, now recast as the prudent ruler, understands how appearances work and exploits them deliberately. Here, hustling becomes *rationalized.* It is no longer vice, but prudence under hostile conditions. Trust becomes dangerous and sincerity a liability. Agency requires strategic distance from moral innocence. Yet Machiavelli still believes in escape. The prince who understands the game can master it. Knowledge restores control. The hustler/sucker distinction is asymmetrical but resolvable. One can learn to hustle.

With Hobbes, the problem deepens. In his *Leviathan*, trust is no longer just some moral disposition (Hobbes 1651/1996). It looks more like what today is referred to as a coordination problem. In the state of nature, trusting others is irrational precisely because others may exploit that trust. The sucker is strategically dominated. Hobbes' solution is institutional. Sovereign power enforces cooperation and removes the incentive to hustle. The tragedy is postponed by authority. Trust is no longer grounded in virtue, but in fear of punishment.

Hume radicalizes this insight. Cooperation depends on conventions that are inherently fragile (Hume 1739/1978). Each individual has an incentive to defect if others comply. The sucker is the cooperator who sustains the convention and the hustler is the free-rider. Importantly, Hume recognizes that cooperation persists not because exploitation is impossible, but because *beliefs about others' behavior remain aligned.* Still, Hume retains optimism. Conventions can stabilize. Education and habit can sustain trust. The tragic horizon remains distant.

Kant represents a return to moral rigor—and a crucial tension (Kant 1785/2012). Kantian ethics demands sincerity, truthfulness, and respect for persons regardless of consequences. From a Kantian perspective, hustling is morally impermissible, even if it is strategically effective. Yet this moral purity comes at a cost. Kant is sometimes criticized for leaving agents strategically exposed (Espíndola & Ramírez 2025). In a world of manipulators, the Kantian agent risks becoming a sucker. Kant's response is rather uncompromising as moral law must not bend to strategic threat. This stance preserves moral dignity, but at the cost of practical vulnerability. His framework contains the seeds of tragedy because it demands innocence in a world increasingly structured to exploit it.

Nietzsche and Marx both reject moralized accounts of exploitation, but in different ways. Nietzsche treats the hustler/sucker distinction as an expression of power (Nietzsche, 1844-1900/2006). Moral norms themselves are tools of the weak, while the strong operate beyond them. Suckerhood is not ignorance, but resentment. Hustling is not strategy, but vitality. Marx, by contrast, shifts the focus to *structures* (Marx 1867). Exploitation is not the result of individual hustlers, but of systemic relations of production. Workers are "suckers" not because they trust capitalists, but because they must sell labor to survive. Hustlers are not clever individuals, but positions within capital. Both thinkers undermine moral blame. But both still hold out hope: revolutionary transformation, revaluation of values, or class consciousness promises escape. What they underestimate is the *epistemic dimension* that will later dominate modern agency.

More than hinted throughout *Sucker Nation*, the momentous philosophical shift occurs with the rise of game theory, social epistemology, and mass society. Thinkers such as (Keynes 1936/2018, Schelling (Schelling 1990) and of course ultimately Aumann show that exploitation can persist even among apparently rational agents, not because of false belief, but because of failures of common knowledge.

Aumann's agreeing-to-disagree theorem now formalizes the devastating insight: If rational agents truly share common knowledge, certain disagreements and exploitative equilibria disappear. Conversely, their persistence signals epistemic fragmentation. The hustler/sucker distinction becomes *structural and inescapable.*

So, the history of philosophy reveals a slow transformation of the hustler/sucker distinction. What begins as a moral error becomes a strategic dilemma, then a coordination problem, and finally a tragic condition of agency itself. What a huge shame for even little scams—and for science itself.

8.2 Scams as Micro-Tragedies

Scams, and online scams, are usually treated as marginal phenomena. Some crude deceptions practiced by criminals against the gullible. News reports, consumer advice, and cybersecurity manuals frame scams as problems of individual vigilance—*don't click, don't trust, be smarter.* Those who fall for scams are careless, naïve, or cognitively deficient while those who avoid them are savvy.

Again this framing is fundamentally misleading in the current framework. Online scams are not just peripheral pathologies of digital life. They are condensed expressions of its central structure. They reveal, in stark and simplified form, the same epistemic and strategic dynamics that govern markets, platforms, and political communication. In particular, they exemplify the hustler/sucker distinction as a stable interaction under epistemic asymmetry and non–common knowledge.

Online scams matter philosophically because they show how agency itself becomes exploitable under conditions of mediated trust, and why awareness alone does not protect agents from loss. They are not merely about deception; they are once more about coordination failure, higher-order belief, and the tragic cost of participation in digital society.

At first glance, scams appear to be straightforward cases of deception. The scammer lies and the victim believes the lie. Remove the false belief, and the problem disappears. This is why most anti-scam discourse focuses on information. Educating users, warning them about tactics and exposing red flags. Yet this model fails to explain several persistent features of online scams (Kipngetich & Ochieng 2025):

- Many victims suspect something is wrong but proceed anyway.
- Many scams are widely known, even clichéd, yet still effective.
- Scam techniques evolve to manage timing, escalation, and commitment.

These facts suggest that scams are not sustained by ignorance alone. They are sustained by *epistemic pressure on agency*: Situations in which acting requires trust, responsiveness, and exposure, while refusing to act carries immediate social or practical costs.

The sucker is the agent who continues to act *as if* certain background norms hold—norms that are necessary for agency itself. Online scams exploit precisely these norms

- that messages are sent by identifiable agents,
- that requests are made in good faith,
- that urgency signals importance,
- that responsiveness is socially expected.

The victim of a phishing email, a romance scam, or a fake support call is often not fully convinced of the scammer's story. Rather, they face a familiar dilemma: *What if this is real?* Ignoring the message risks loss (account access, relationship, reputation, riches easily made) yet responding risks exploitation.

The victim's action may more often than not still be subjectively rational. Given uncertainty, given asymmetric costs of error, and given the expectation that most interactions are legitimate, engagement becomes the safer option—until it is not. The sucker is the one who cannot afford not to believe enough to act.

The online scammer—the hustler—rarely relies on a single lie. Instead, they engage in epistemic engineering by shaping the victim's beliefs about time, risk, and coordination. Common scam techniques illustrate this clearly (Akerlof & Shiller 2015, Dove 2021):

- *Phishing and Authority Scams*: Emails purporting to be from banks, employers, or platforms rely less on plausibility than on *institutional asymmetry*. The victim knows the message *might* be fake, but also knows that failing to respond to legitimate institutional communication can be costly. The scam exploits uncertainty about *procedural norms*, not belief in content.

- *Romance Scams*: These involve extended interaction, gradual trust-building, and emotional investment. The hustler does not rush deception. They allow the victim to *commit themselves epistemically and affectively*. By the time money is requested, the cost of disbelief has risen dramatically—not because the victim is convinced, but because withdrawal would invalidate their prior investment.

- *Tech Support Scams*: Here, hustlers exploit epistemic dependence. Victims often know they lack technical understanding. The scammer's advantage lies not in superior knowledge, but in *controlling the frame* in which knowledge claims are made. The victim defers not because they believe blindly, but because they cannot independently verify in time.

In each case, hustling consists in managing what the victim believes about their own uncertainty, about institutional expectations, and about the costs of delay.

Scam warnings circulate constantly. Memes mock obvious frauds. Yet scams adapt and continue. This persistence mirrors the broader hustler/sucker equilibrium. Awareness fails because it does not generate common knowledge at the moment of action. The victim may know scams exist, but they do not know whether this specific interaction is one. They may know others are cautious, but they do not know whether institutional procedures assume responsiveness. Moreover, refusing to act unilaterally carries costs. Ignoring emails, messages, or calls systematically is not viable in a digital society. Agency requires engagement. The scammer exploits this necessity. Online scams persist because digital participation structurally exposes agents to asymmetric risk.

Scams often proceed through stages, each requiring small acts of compliance. Replying to an email, clicking a link, sharing limited information. Each step is individually rational and low-risk. Together, they create *lock-in.* This mirrors the tragic structure of larger hustler/sucker equilibria:

- Early participation is low cost and rational.
- Exit becomes increasingly costly over time.
- Insight often arrives only after commitment.

Victims frequently report the same experience—something like: "I realized something was wrong, but by then I had already...". This is *path dependence under uncertainty.* The tragedy lies in the fact that agency itself produces vulnerability over time. Acting reasonably at each step leads to cumulative exposure.

After a scam, victims often experience intense shame (Balcombe 2025). They hesitate to report the incident or seek help. This silence is not accidental—it is part of the system. Shame reinforces the hustler/sucker structure by isolating victims, preventing collective learning and preserving epistemic asymmetry. In this sense, scams extend beyond the initial interaction. They generate *secondary suckerhood*, where victims bear the cost of exposure twice. First financially or emotionally. Then socially. This parallels broader digital dynamics, where vulnerability is moralized and failure individualized. The victim is told they should have known better, reinforcing the illusion that hustling is avoidable through intelligence alone.

Seen through the developed framework, online scams are *micro-tragedies.* They compress into individual interactions the same structural features that characterize modern agency at scale trust as a precondition of action, trust as exploitable, knowledge as insufficient for escape and exit as costly without coordination. Academia has similar micro-tragic stories to tell.

8.3 Academic Hustlers and Suckers

Contemporary academia is characterized by a paradox that is no longer controversial but still unresolved. There is the exponential growth of publications alongside a widespread conviction that much of this output advances neither understanding nor truth. Replication crises, metric gaming, fashionable theorizing, and incremental "least publishable units" are openly discussed within the academy itself (Waterman et al. 2023). Awareness is high while reform is low. This persistence reflects a stable hustler/sucker equilibrium embedded in a *publish-or-perish* regime (Nature 2010) governed by attention economics. Scholars act rationally given the incentives they face, even when those incentives conflict with the epistemic ideals that historically justified science as a social institution.

Sociologist, Robert K. Merton, famously articulated the normative ethos of science as CUDOS: Communism (shared knowledge); Universalism; Disinterestedness and Organized Skepticism (Merton, 1973). These norms presuppose that recognition follows epistemic contribution and that priority serves knowledge rather than visibility. What has changed is not that these norms are rejected, but that they have been structurally inverted. Recognition now precedes evaluation as much as visibility precedes validation. Priority is no longer secured by contribution alone, but by speed, branding, and strategic positioning.

Merton already noted the tension between priority, reward and competition. What the attention economy adds is a new conversion mechanism attention substitutes for epistemic judgment, while preserving the symbolic language of scientific norms. The result is not norm violation, but norm simulation.

"Publish or perish" may be understood as a coordination failure under non-common knowledge. Individual scholars may prefer slower, riskier, deeper work aligned with Mertonian ideals. Yet because hiring, tenure, and funding depend on output proxies, unilateral restraint is professionally suicidal. Each academic knows the system is distorted. What they do not know is whether others are willing to act on that knowledge. Without coordination, awareness becomes inert. Complaints keep circulating although behavior does not change—it just becomes a matter of *Winning the Game Scientists Play* (Sindermann 2008).

The *academic sucker* is the sincere scholar who continues to publish *as if* publication were primarily about advancing knowledge, while knowing that the system rewards visibility, speed, and conformity. Suckers of the academy trade epistemic wits—depth, patience, risk—for academic vanity like legibility, continuity, and survival. Now, the *academic hustler* understands how attention converts into prestige and how prestige converts into power. Hustlers brand concepts, dominate citation networks, set agendas, and control gatekeeping institutions much in the spirit of the "Matthew's Effect in Science" also coined by Merton (1968). Their advantage lies not in violating Mertonian norms explicitly, but in exploiting their symbolic persistence. And just as before, hustlers depend on suckers. The system requires many sincere contributors for a few to extract attention rents. Exploitation persists without deception.

In Merton's realm of thought, vanity is disciplined by communal norms. In contemporary academia, vanity has been structurally redefined. As Franck argued, attention functions as capital (Franck, 2020). Before a contribution can count as knowledge, it must be seen, cited, indexed, and ranked. Publishing becomes attention labor. Titles are optimized, novelty overstated, keywords curated, and citations managed strategically—not out of narcissism, but necessity. Vanity is no longer opposed to knowledge but rather the infrastructure through which knowledge must pass to exist institutionally.

Metric governance intensifies this inversion. Quantification promises objectivity but replaces judgment with proxies (Muller, 2018). Once the case that proxies govern rewards, agents rationally optimize for them. This is the epistemic version of *Goodhart's Law*: "When metrics become targets, they cease to measure scientific quality" (Goodhart 1975). Mertonian norms remain rhetorically intact, but practically sidelined. Transparency increases sophistication but does not add up to resistance. By the end of the day, irony replaces reform.

And now add AI into the mix. AI-assisted writing marks a qualitative shift because it fits the existing equilibrium. Large language models excel at producing attention-legible text in terms of literature reviews, incremental contributions, stylistic conformity and citation-shaped prose (Cooperman & Brandão 2024). For the academic sucker, AI becomes a defensive necessity meeting output demands while preserving scarce cognitive resources. Refusal risks professional disadvantage. Yet, for the academic hustler, AI is a force multiplier—accelerating agenda-setting and amplifying visibility dominance. Control over attention, not authorship, becomes pivotal.

In and by itself, AI does not corrupt a healthy system. It reveals what the system already rewards as to form over contribution, circulation over skepticism and output over judgment. AI makes explicit what was previously implicit in that much academic publishing is no longer primarily about advancing shared knowledge in the Mertonian sense, but about maintaining position within an attention economy. The distinction between genuine inquiry and its simulation becomes harder to enforce, not because norms disappear, but because incentives actively penalize their enforcement. On top of that the hustler/sucker distinction sharpens. AI reallocates costs without changing the structure. The sucker bears the burden of sincerity yet the hustler captures scale.

Once more from the tragic perspective, there is no innocent exit. To stop publishing is to abandon agency. To continue—especially with AI assistance—is to accept further separation between visibility and knowledge. Wisdom clarifies the damage but it does not undo it. Reform requires coordinated institutional change—evaluation criteria, hiring norms, funding structures—not individual virtue. Without common knowledge and synchronized restraint, appeals to integrity merely redistribute risk.

8.4 Toy of Tragedy

This is the situation we are seemingly faced with these days: In information environments that reward attention rather than accuracy, even fully rational agents, updating their beliefs correctly, may systematically converge on false, polarized, and unstable beliefs. Individual rationality intact, no stupidity, no bad faith and no ignorance required. Everyone, reasons correctly, updates evidence properly and acts in their own epistemic interest. And yet things go wrong.

The breakdown comes from selective exposure, feedback loops, asymmetric visibility and amplification of extreme signals. Truth loses not because it is false, but because it is *less attention-efficient.* From here polarization for instance becomes *inevitable*, not accidental. Once the system has algorithmic filtering, engagement incentives and social reinforcement, then disagreement increases with information and corrections backfire, confidence grows while accuracy falls. This reverses the classical Enlightenment assumption that more information leads to consensus. The same structure and dynamics seem to explain market bubbles, fake news, conspiracy theories, strategies of scientists and democratic erosion. Truth is not defeated by irrational people, but by rational people placed in the wrong information system.

For a toy model of tragedy, specify world, hypotheses and agents accordingly:

- Binary state of the world: $\theta \in \{0, 1\}$.
- n agents $i = 1, \ldots, n$.
- Each agent has a prior $p_i^0 = \Pr(\theta = 1)$.

Reliably derived signals, read truth, are available, but not equally surfaced. At each discrete time $t = 1, 2, \ldots$, agent i could receive a signal $s_{i,t} \in \{+1, -1\}$ the reliability of which depends on the state of the world θ:

$$\Pr\left(s_{i,t} = +1 \mid \theta = 1\right) = q,$$

$$\Pr(s_{i,t} = +1 \mid \theta = 0) = 1 - q$$

with $q \in (1/2, 1)$. So signals are genuinely informative.

Instead of receiving an unbiased draw from the signal stream, a social platform, say, shows agent i a *selected* signal. Let $x_{i,t} \in \{+1, -1\}$ be what agent i actually sees. Define agent i's current belief in log-odds:

$$l_{i,t} = \log \frac{p_{i,t}}{1 - p_{i,t}}.$$

- $p_{i,t}$ is agent i's subjective probability, at time t, that some proposition or event is true (for example, that a claim is correct, a project will succeed, or a signal is reliable).
- The fraction $\frac{p_{i,t}}{1-p_{i,t}}$ is the odds in favor of the event.
- Taking the logarithm of those odds converts the belief into *log-odds*, denoted $l_{i,t}$.

So $l_{i,t}$ measures how strongly agent i believes the proposition, on a symmetric scale:

- $l_{i,t} = 0$ means the agent assigns probability ½ (complete uncertainty).
- Positive values mean the agent believes the proposition is more likely true than false.
- Negative values mean the agent believes it is more likely false than true.

- Large absolute values mean strong confidence.

The advantage of expressing beliefs in log-odds is that updates become additive as new evidence shifts $l_{i,t}$ up or down by a fixed amount, rather than requiring nonlinear probability calculations. Log-odds turn belief updating into a process of accumulating reasons for or against a claim. In short, $l_{i,t}$ is a numerical measure of *how convinced agent i currently is*, expressed in a way that makes belief change easy to track and compare over time.

Suppose for sake of the argument that the social platform selects content to maximize predicted engagement, which increases when content *matches* the user's current leaning (confirmation/affect alignment). Introduce a simple parametric selection rule conveying that agent i's action at time t is chosen probabilistically, with the probability depending on how strong their current belief is:

$$\Pr(x_{i,t} = +1 \mid s_{i,t} = +1, l_{i,t}) = \sigma(\alpha\, l_{i,t}),$$

$$\Pr(x_{i,t} = -1 \mid s_{i,t} = -1, l_{i,t}) = \sigma(-\alpha\, l_{i,t}),$$

where

$$\sigma(z) = \frac{1}{1 + e^{-z}}$$

and $\alpha \geq 0$ measures the strength of attention-optimization.

- If $\alpha = 0$: No filtering (neutral feed).

- If $\alpha > 0$: The feed increasingly serves belief-consistent items as confidence grows.

Here is the intuition step by step:

- $x_{i,t} \in \{+1, -1\}$ is the *action* or *expressed choice* of agent i at time t (for example, endorsing vs. rejecting a claim).
- $s_{i,t} \in \{+1, -1\}$ is the *signal* the agent receives at time t (e.g. favorable or unfavorable evidence).
- $l_{i,t}$ is the agent's *current belief* expressed in log-odds.
- $\sigma(\cdot)$ is a *sigmoid function*, which smoothly maps real numbers into probabilities between 0 and 1.
- $\alpha > 0$ is a *sensitivity parameter* that captures how strongly belief influences action.

The first line of the parametric selection rule

$$\Pr(x_{i,t} = +1 \mid s_{i,t} = +1, l_{i,t}) = \sigma(\alpha\, l_{i,t})$$

means: If the agent receives a positive signal, the probability that he/she chooses the positive action increases with how strongly the agent in question already believes the proposition. The stronger the belief (higher $l_{i,t}$), the more likely the agent is to act in line with that belief—but never with certainty. The second line of the parametric selection rule

$$\Pr(x_{i,t} = -1 \mid s_{i,t} = -1, l_{i,t}) = \sigma(-\alpha\, l_{i,t}),$$

means: If the agent receives a negative signal, the probability that he/she chooses the negative action increases the more strongly the agent's belief favors rejection. Taken together, these equations describe a *soft, probabilistic decision rule*:

- Agents are more likely to act in accordance with their beliefs,
- but they sometimes deviate, especially when beliefs are weak,
- and the parameter α controls how deterministic or noisy behavior is.

Beliefs *tilt* behavior, they do not dictate it. Strong beliefs make aligned actions very likely; weak beliefs make actions closer to random. This is a tilted likelihood in the sense that attention acts like an extra, belief-dependent bias on what evidence is encountered.

Next, agents update by Bayes' rule using the likelihood of the observed item $x_{i,t}$. In log-odds form:

$$l_{i,t+1} = l_{i,t} + \log \frac{\Pr(x_{i,t} \mid \theta = 1)}{\Pr(x_{i,t} \mid \theta = 0)}.$$

If agents *naïvely* treat the feed as an unbiased sample of the world (they don't model the selection mechanism), then the increment is the usual constant step:

$$l_{i,t+1} = l_{i,t} + x_{i,t} \cdot \lambda, \lambda = \log \frac{q}{1-q}.$$

This may even be strengthend because even if they *do* model the selection but can't observe the hidden counterfactuals, the posterior can still become brittle under plausible identification limits; but the cleanest toy result uses the naïve sampling assumption.

Define the *expected drift* in log-odds conditional on current belief l:

$$\Delta(l) = E[l_{t+1} - l_t \mid l_t = l, \theta].$$

$\Delta(l)$ measures how an agent's belief is expected to change, on average, given their current level of belief. More specifically:

- l_t is the agent's *belief* at time t, expressed in log-odds.
- $l_{t+1} - l_t$ is the *change in belief* from one time step to the next.
- The *expectation* $E[\cdot]$ means averaging over all the randomness in what happens next (new signals, actions, noise, etc.).
- The condition $l_t = l$ means *fixing* the agent's current belief at the value l.
- θ again denotes the underlying state of the world or true parameter that governs how informative signals really are.

The expression $\Delta(l)$ thus answers the question: If an agent currently holds belief l, how should we expect that belief to move, on average, in the next step?

- If $\Delta(l) > 0$, then an agent with belief l is expected to become *more confident* (log-odds increase).

- If $\Delta(l) < 0$, they are expected to become *less confident* (log-odds decrease).
- If $\Delta(l) = 0$, belief l is, on average, *stable.*

In other words, $\Delta(l)$ describes the directional "pull" on beliefs at level l whether beliefs tend to drift upward, downward, or remain stationary when they are currently at l.

Finally back to *tilting.* Under the tilted-evidence rule, for $\alpha > 0$ the probability of seeing confirmatory evidence becomes increasing in $[l]$. This creates *self-reinforcement*:

- When $l > 0$, $\Pr(x = +1 \mid l,\theta)$ increases with l, so $\Delta(l)$ becomes more positive as l grows.
- When $l < 0$, $\Pr(x = -1 \mid l, \theta)$ increases in magnitude with $[l]$, so $\Delta(l)$ becomes more negative as l decreases.

So $l = 0$ becomes an unstable fixed point and the dynamics tend toward *absorbing extremes* (very large $[l]$), i.e. polarization.

Even if the raw signal process favors the true θ, the selection bias can dominate. In particular, for sufficiently large α, the platform-induced tilt can make

$$E[x_t \mid \theta = 1, l \ll 0] \approx -1,$$

so an initially skeptical agent keeps seeing "evidence" against the truth, updates rationally on it, and becomes more confident in the false view. If the true state of the world is positive ($\theta = 1$), but the agent's current belief is very strongly negative, then the agent is expected to choose the negative action almost all the time. In unpacked detail:

- $\theta = 1$ means that, objectively, the *proposition is true* or the positive state holds.
- $l \ll 0$ means the agent's belief (in log-odds) is very negative—the agent is *strongly convinced* that the proposition is false.
- $x_t \in \{+1, -1\}$ is the agent's *action or expressed stance.*
- $E[x_t \mid \cdot]$ is the average (*expected*) action under those conditions.
- ≈ -1 means the action will be *negative with very high probability.*

Even though the world is actually in the positive state, an agent who is deeply convinced of the opposite will almost certainly act as if the negative state were true. That's the formal sense in which

$$\text{(rational updating} \wedge \text{distorted exposure)} \Rightarrow \text{systematic error.}$$

That's basically formal bubble formation. That's truly bad (formal) news for modern agency.

8.5 Modern Agency

Modern agency is often imagined as empowered, reflexive, self-directing. Yet this image rests on a background assumption that no longer holds. That appearing, speaking, publishing and participating are acts whose meaning is broadly shared and whose interpretation is relatively stable. Appearance no longer guarantees recognition as little as expression guarantees interpretation. What appears is processed, counted, ranked, and monetized. Meaning is not denied, but it is no longer central to how social systems respond.

This transformation does not abolish agency, but it does make it tragic. To act as a social agent—to speak, to signal, to cooperate—one must behave *as if* meaning still attaches to appearance, even while knowing that it often does not. Agency therefore requires a *necessary misrecognition.* One must believe enough to act, but not so much as to be destroyed by that very belief. This is the tragic starting point since *agency demands trust in conditions that systematically betray trust.*

The sucker is the agent who *bears the cost of social intelligibility.* Suckers trust where trust is required for coordination. They participate where participation is required for presence. They expose themselves where exposure is required to have a face at all. The sucker's defining feature is fidelity to the background norms that make interaction possible. They behave *as if* sincerity matters, *as if* cooperation is binding and *as if* participation is meaningful. These are the presuppositions of social life itself. And tragedy enters because these presuppositions are exploitable. The sucker loses because they reason in a way that makes agency possible for everyone—including those who will exploit it. They pay the price of meaning so that meaning can exist at all. Suckerhood is not an error to be corrected. It is a role demanded by the structure of the game.

Opposite the sucker stands the hustler. The hustler may be unscrupulous but more to the point they are *epistemically awakened.* They see what the sucker does not—or rather, what the sucker must partially deny in order to act. They see that sincerity is legible only as signal, that trust is not reciprocated by default, that visibility is not recognition but extraction. The hustlers maintain separation between belief and expression, identity and appearance, participation and commitment. They know the game is not what it presents itself to be. Yet this knowledge is not free. The hustler pays in alienation. They cannot fully inhabit the norms that make social life feel meaningful. They live in irony, calculation, and strategic restraint. They see through the game, but cannot exit it without forfeiting influence, presence, or livelihood. The tragic structure asserts itself therein *knowledge itself becomes a form of loss.* The hustler survives by refusing innocence, but that refusal corrodes trust, belonging, and shared meaning.

The tragic exchange at the heart of modern agency is that *wits are traded for vanity.* Vanity is the belief that one's appearance matters as appearance—that being seen is being recognized, that being heard is being understood, that being authentic is being protected. This belief is historically grounded in social forms where it was largely true.

In modern attention economies vanity becomes a trap. Visibility is rewarded, but not for its meaning. Authenticity is encouraged, but only because it allocates attention and produces legible data. Outrage spreads, but not because it persuades. The system does not care what one means—only that one appears. To refuse this trade—to insist on wits over vanity—is to accept invisibility, marginality, or irrelevance. To accept the trade is to accept exposure, predictability and exploitation. There is no option that preserves both strategic intelligence and social presence intact.

Game theory sharpens this tragedy by showing why insight does not dissolve it. Hustlers and suckers persist because the relevant knowledge is not common knowledge. Without common knowledge, coordination fails. Each agent continues to play their role because deviating alone appears worse than continuing together. The sucker continues to trust because withdrawal would be isolating. The hustler continues to exploit because exploitation remains profitable. The equilibrium holds because no one can escape it unilaterally.

Under common knowledge of rationality, this equilibrium would collapse. Sucker strategies would no longer be rationalizable and hustler strategies would lose their foundation. But common knowledge is extraordinarily rare in modern societies. Scale, mediation, and algorithmic opacity ensure that higher-order beliefs remain misaligned. The tragedy deepens because even knowing the structure does not free the agent from it.

It may even be argued that tragedy is not only epistemic but also affective. Shame plays the role of fate. Once one has exposed oneself—taken a public stance, signaled identity, invested authenticity—withdrawal becomes costly. Silence feels like defeat. Revision feels like hypocrisy. Distance feels like betrayal of one's earlier self. Thus realization often arrives after commitment, and commitment binds the agent to continued participation. The system does not merely extract attention. It extracts *identity continuity*, making exit feel like self-annihilation. Time compounds the tragedy. Many accept suckerhood as apprenticeship: *I will endure exposure now so that I may gain autonomy later*. Most never do. Those who succeed often discover that success requires perpetuating the same structure for others. Tragedy without climax—a slow accumulation of loss without catharsis.

Platform society often enough sanctify suckerhood. Exposure is praised as courage. Authenticity is moralized. Vulnerability is celebrated. But strategic restraint is condemned as cynicism or bad faith. This moral inversion stabilizes the tragic structure. It ensures that those who bear the cost of meaning are praised for their exploitation. It discourages strategic learning yet stigmatizes otherwise healthy distance. The sucker becomes virtuous when the hustler becomes unspeakable. Yet the system rewards the latter and consumes the former. This is indeed tragic irony in the classical sense: Virtue produces vulnerability, and vulnerability produces loss.

Classical tragedy sometimes ends in death, craziness,

exile, or divine judgment. Modern tragedy offers no such release. There is no *deus ex machina* that restores innocence, no revelation that dissolves the conflict. The hustler/sucker distinction cannot be abolished without abolishing trust, meaning, and social life itself. Nor can it be resolved by moral reform or individual intelligence. It is a permanent tension built into the conditions of modern agency. What to do about it?

CHAPTER 9: Escaping the Trap

Any serious conclusion to the hustler/sucker framework must begin by resisting consolation. The temptation to offer solutions—to provide steps, strategies, or prescriptions—misunderstands the nature of the problem. The hustler/sucker distinction is a structural feature of modern agency under conditions of epistemic opacity, attention scarcity, and compulsory visibility. To promise a clean escape would be to reproduce the very illusion that sustains suckerhood along with the the belief that insight alone restores control.

This conclusion therefore does not close the argument by resolving it. It closes it by clarifying what kind of agency remains possible once resolution is ruled out. It treats escape not as liberation from the tragic condition, but as movement within it—movement that accepts loss, renounces innocence, and refuses both naïveté and cynicism as stable endpoints. Modern agency cannot be restored in pure form but it can be re-composed under damage.

V. F. Hendricks, *Sucker Nation*,
https://doi.org/10.1007/978-3-032-22788-1_9

9.1 One Last Time

At the heart of the framework lies the deceptively simple distinction. The sucker is the agent who behaves as if trust, sincerity, participation, and authenticity still function as they once did—*as if* appearance were recognition, expression were interpretation, and cooperation were reciprocated. The hustler is the agent who knows this is no longer the case and who exploits the gap. The distinction is epistemic. The sucker is subjectively rational under beliefs that make agency at all possible. The hustler is not immoral but merely strategically rational under conditions of belief asymmetry. Both are responses to the same environment. The hustler/sucker distinction is an equilibrium under epistemic fragmentation.

It would misunderstanding the framework to just treat escape as simply "not being a sucker." Surely, such a framing is seductive but wrong. To "not be a sucker" habitually means not trusting, not exposing, not participating and not believing in the sincerity of others. Behaviors do not restore agency. They undermine its conditions. Trust is a precondition for coordination. Exposure is not merely vanity but typically the price of having a voice. Participation is how collective action becomes possible at all. Eliminating these wholesale does not produce autonomy. In matters of deliberation, decision and action, it produces paralysis, isolation, or irrelevance. Moreover, universal refusal of suckerhood is incoherent. If no one were willing to trust, cooperate, or expose themselves, social life would collapse. The hustler/sucker distinction exists precisely because some degree of suckerhood is functionally necessary. Thus escape means reallocating vulnerability.

If not being a sucker fails, perhaps becoming a hustler succeeds. This, too, is an illusion. The hustler appears free because they preserve distance. They do not collapse belief into expression. They do not equate authenticity with sincerity. They do not assume norms are binding. They treat visibility strategically and extract value where others expose themselves. Yet the hustler is not sovereign. They are dependent on the very norms they disavow. Hustlers require suckers in order to exist. They rely on the continued participation of those who still believe enough to expose themselves. Their advantage is parasitic. Add to this that hustling extracts its own price. It corrodes trust not only in others, but in the possibility of shared meaning. It replaces participation with calculation, belonging with leverage, and speech with signaling. The hustler may avoid being exploited, but they also avoid being fully committed. Hustling is not escape either. It is strategic accommodation within the same tragic field.

Then how about forgetting the whole thing? Total withdrawl is tempting. Silence—invisibility—refusal. If appearance is monetized anyway and everything is a hustle, perhaps the solution is to disappear. Withdrawal can be protective, and in limited contexts it is often even necessary. But as a general strategy, it fails to restore agency because agency requires appearance. To act politically, socially, culturally or economically is to be visible to some audience. In technofeudal systems, disappearance does not disrupt extraction but removes one from relevance. The field remains, populated by those most willing to exploit it. Moreover, withdrawal is rarely interpretable as resistance unless it is coordinated. Without common knowledge, silence looks like absence, not dissent. The withdrawing agent does not change the game. They exit it alone. Thus withdrawal preserves integrity at the cost of agency. It is not liberation but closer to tragic resignation.

9.2 Reframing Escape

As these exits are rejected, escape must be re-conceived. Escape is acting lucidly within the tragic condition. In classical tragedy, the hero does not overcome fate. They come to understand it. Their final dignity lies not in victory, but in clarity—clarity that arrives too late to prevent loss, but not too late to prevent self-deception—Camus' *The Myth of Sisyphus* (1942) comes to mind.

Similarly, modern agency can only be regained tragically. That is, it can only be regained by abandoning the fantasy of innocence, refusing the fantasy of mastery while acting under acknowledged loss or sunk cost. Escape is not transcendence but recomposition. Agency is to be regained by choosing it.

The sucker is vulnerable everywhere, all the time, by default. The hustler is vulnerable nowhere, except where strategically profitable. Regained agency lies in neither extreme. *Selective vulnerability* essentially amounts to exposing oneself where exposure enables genuine action, withholding exposure where it merely feeds extraction, trusting where trust can be reciprocated and finally by refusing trust where it is predictably exploited. It sounds like moral purity but is actually about damage control. Just because some suckerhood is unavoidable does not entail it should be universalized.

One of the deepest losses in modern agency is the collapse of epistemic distance—the forced alignment of belief, identity, and expression. Platform society pretty much demands that agents be fully legible at all times. It treats interiority as suspicious and restraint as bad faith. Regaining agency requires restoring distance. Distance means not treating every belief as an obligation to speak or perform, not binding one's public identity to every moral impulse and not allowing every reaction to become a signal.

The sucker collapses inner life into outer display. The hustler instrumentalizes that collapse. Agency lies in preserving interiority without withdrawing from action. This distance is fragile and socially costly. But without it, agency collapses into performance. In worst case we will be forever "amusing ourselves to death" (Postman 1985).

Silence plays a significant role in regained agency, but only when properly understood. In attention economies, reaction is currency. Outrage, correction, amplification—all pay rent. Silence interrupts this very flow. It is not absence, but refusal to be extracted from. Now, silence becomes agency only when it is selective rather than total, when patterned rather than sporadic—when it is socially intelligible. Thus silence is in essence withdrawal from automatic participation. If you don't have something qualified to say, don't say anything—anything else is yet again just feeding the beast of suckerhood.

One of the most corrosive illusions of modern life is that agency is purely individual and "What About Me Ism" (Hendricks 2025). The sucker nation exposes this illusion decisively. Exploitation persists because they lack coordination. Without common knowledge, unilateral refusal fails. Each agent waits for others to move first. The equilibrium holds. So regained agency therefore cannot be purely individual. It requires local coordination in small groups bounded publics with shared norms and repeated interactions. These contexts reduce epistemic opacity. They make belief alignment possible. They allow trust to function without immediate exploitation.

Regaining agency is about abandoning moralized off the shelf explanations and instead focus on the epistemic conditions of agent interactions. Suckers are not simply foolish. Hustlers are not just evil. Both are responding rationally to structural constraints. Understanding this shifts the burden from individuals to systems. It replaces shame with clarity. It reveals that much of what feels like personal failure is ontological indebtedness—the requirement to pay rent on one's own face in order to exist socially at all. This structural humility clarifies its limits.

Regained agency does not look anything like heroic unfortunately. It looks modest partial and somewhat damaged. It looks like

1. *acting without expecting recognition,*
2. *trusting without expecting safety,*
3. *speaking without expecting sincerity to protect meaning* and
4. *withdrawing without expecting victory.*

It looks like choosing when to appear and when not to, without mistaking either choice for purity.

9.3 Platform Support?

Social platforms have so far been cast and characterized as central villains of contemporary social life. They monetize attention, reward exposure, flatten meaning, and turn sincerity into extractable signal. Social platforms appear as hustler mentality permeating through technofeudal infrastructures. As oxymoronic as it sounds, privately owned public spaces of appearance in which agents must pay rent on their own face simply to exist socially at all. Within this diagnosis, platforms entrench the hustler/sucker distinction rather than offer any path beyond it.

And yet, if the tragic framing of modern agency is taken to heart, the conclusion stands to be refined. The tragic claim is that agency now exists only in damaged, mediated, and recomposed form. If there is no exit from mediation, then the central question becomes whether social platforms—despite and because of their pathologies—

can support limited, partial, but real forms of agency that do not collapse entirely into either suckerhood or hustling.

If they were to choose to, social platforms may indeed aid in recomposing agency by systemically enabling selectivity, bounded coordination, and epistemic distance. Platforms do not liberate agency but they may provide essential materials and conditions from which damaged agency can be reassembled.

Many critiques of social platforms frequently presuppose a sovereign model of agency. A subject who could act freely, transparently, and meaningfully if only manipulation, misinformation, or surveillance were removed. From this perspective, platforms can only appear as obstacles or corruptions. Now, the tragic framework rejects this premise. Agency today is irreducibly mediated. It unfolds under conditions of epistemic opacity, attention scarcity, and compulsory visibility. Trust is exploitable, participation extractive, and authenticity monetized. These conditions cannot be undone by better design or individual enlightenment alone at least not under the current business model of social platforms (Hendricks & Mehlsen 2022).

Recomposition means rearranging the compromised elements like visibility, trust, participation and silence into configurations that preserve some capacity to act, even while acknowledging permanent loss. It is agency without purity, freedom without sovereignty. Social platforms must be evaluated not by whether they restore ideal agency, but by whether they allow agents to recompose agency strategically within damaged conditions.

One of the core mechanisms sustaining suckerhood is indiscriminate visibility—no backstage, only front stage all the time. The suckers expose themselves continuously, sincerely, and universally, believing that appearance will be met as recognition. The hustler exploits this exposure, extracting value from legibility. Social platforms intensify this dynamic by default. They encourage constant posting, immediate reaction, and full legibility. But they also—often inadvertently—provide tools for indeed quite selective visibility.

Agents may actually choose when to post and when not to, segment audiences, maintain multiple registers of expression, engage intermittently rather than continuously. Obviously, used naïvely, these affordances merely deepen extraction. Used tragically—that is, without belief in innocent participation—they allow agents to decouple presence from total exposure. One can appear without fully revealing, speak without fully committing, and participate without surrendering interiority. This selectivity does not abolish rent extraction. But it reduce the scope of what is extractable. Agency is not regained by refusing visibility altogether, but by rather by controlling its distribution.

A central insight of the hustler/sucker architecture is that exploitation persists because rationality, intentions, and vulnerability are not common knowledge. Large and largely anonymous publics generate epistemic fog. Agents know much, but do not know what others know, or whether withdrawal would be coordinated. As a result, sucker strategies persist even among the informed. Platforms exacerbate this fog at scale—but they also enable *bounded publics*. Groups, sub-communities, and semi-stable networks allow for repeated interaction, shared expectations, and partial belief alignment. These spaces are fragile and exclusionary, but they do matter. Within bounded publics trust becomes locally rational, cooperation becomes sustainable, sincerity regains partial meaning and hustling becomes more detectable and costly. Agency here is not universal or pure. It is *situated*. The agent does not trust everywhere, with everyone. They trust *here*, *now*, *with these others*. This localization of trust does not eliminate suckerhood may confine it. Platforms may accordingly aid agency not by restoring general trust, but by enabling its containment. That's at least a start. Down the line social platforms may even decide to stimulate common knowledge (of rationality) even if it hurts their business model. *Social platforms as responsible civil society actors—wouldn't that be refreshing?*

Attention economies depend on speed. Reaction is rewarded yet hesitation is punished. One reason suckers are exploitable is that they are drawn into continuous responsiveness in terms of liking, sharing, commenting, correcting, reacting—no time for epistemic distance. Platforms intensify this pressure, but they could allow for *temporal agency* in terms of delayed posting, implementing friction strategies, intermittent engagement and asynchronous participation and other frictions strategies giving users the opportunity to reflect. These days, such temporal affordances are either abolished or neglected at the outset—

they are bad for business. They slow down agent interactions, less traffic means less data to harvest about users, less data is less profiling and less credit information about users, diminishing add revenues and cripples prediction capabilities.

Yet the very same temporal affordances are central to recomposed agency. By slowing the conversion of impulse into signal, agents regain a measure of control over how much of their interior life is turned into extractable data. This is not disengagement. It is re-timing participation. It interrupts the automaticity that feeds exploitation and increase the signal to noise ratio.

From the tragic framework it follows that individual enlightenment is insufficient. Hustler/sucker equilibria persist because agents cannot coordinate unilaterally. Platforms, despite their role in epistemic fragmentation, remain among the few infrastructures capable of supporting local coordination. The coordination is modest and non-heroic but may enforce shared norms within communities, explicit agreements about exposure, and collective refusals of participate in certain attention games from subversive challenges. Such platform initiatives would reshape the local game making suckerhood less diffuse and hustling less profitable within these bounded contexts. Platforms will not thus function not as emancipatory tools, but as *coordination technologies*—means by which damaged agents can align action under partial knowledge.

Possibly the most significant contribution of platforms to recomposed agency is that they make innocence untenable. Metrics, algorithms, monetization, and amplification are visible enough that naïve participation becomes difficult to sustain. This loss of innocence is painful. It corrodes trust and undermines meaning. But it also prevents full surrender to vanity. It makes it harder to believe that exposure is unmediated, that authenticity is protected, or that participation is harmless. Agency after innocence is not free. But it is *less easily hijacked.*

In the end, regained agency is not freedom from the hustle. It is *freedom from illusion*—the illusion that one can act without cost, appear without risk, trust without loss, or know without damage. Wisdom here is tragic in kind wisdom. It's clarity without innocence, agency without purity, action without guarantees. To escape the hustler/sucker situation is not to leave it here and now. It is to *inhabit it consciously*, to refuse both naïveté and cynicism as total positions, and to live with the scar rather than pretending it can be healed. Sooner or later we may come to realize that this can of worms only opens from the inside.

Acknowledgements

Sucker Nation: Trading Wits for Vanity is in many ways the intellectual testament to the research conducted at *CIBS: The Center for Information and Bubble Studies* at the University of Copenhagen which I founded and directed from 2015-2023. CIBS was funded by the first Carlsberg Foundation Semper Ardens Grant with supplementary funding coming from TrygFonden.

CIBS was populated by—save for a few suckers—the brilliant brains of good people: From administrative staff, undergraduate and graduate students, PhD-students, postdocs, operatives and contractors from the public and private sector and faculty members ranging from philosophy to semiotics over psychology to mathematics, computer science and economics. A truly interdisciplinary cast of characters, committed, calm, cool and collected. I would like to thank them all for realizing—in ways I could never have imagined—an idea just fostered in my head and then carried out to perfection by the group at CIBS. You made this thing of ours come true.

As to this book in particular, I would like to thank Rasmus K. Rendsvig, Mads Vestergaard & Joachim Wiewiura—

V. F. Hendricks, *Sucker Nation*,
https://doi.org/10.1007/978-3-032-22788-1

all former members of CIBS—for their constructive suggestions and critical comments on earlier drafts of the manuscript.

Great colleagues including professors Gregory Wheeler, Olivier Roy, Anders Rahbek, Carl-Johan Dalgaard, Anders Søgaard and post docs Sandrine R. Schiller and Joshua James Hatherley suggested valuable improvements to some of the arguments presented. Former students, Ditlev Pihl Bisgaard and Mikkel Møller-Sørensen, also provided valuable input on the way.

Students at the University of Copenhagen who—over the past six years—attended my undergraduate course on *(Mis)information and Democracy*, were a great source of inspiration and provided valuable insights and suggestions many of which found their way into the book.

I would like to thank my editor Christopher Wilby for his great help and enthusiastic support all the way to publication. I'm grateful to Ties Nijssen, Floor Oosting, Palani Murugesan and Springer Nature for taking on *Sucker Nation.*

Good friends to the rescue too including Teit Molter and Bjarke Malmstrøm Jensen providing moral support, encouragement and insight—thank you.

The family was along for the ride. My cousin, Nanna Grunwald, came up with the magnificent cover based on a very preliminary sketch and my adoration for wrinkled brown bags. The pride of my life, my son, Milton W. Hendricks chipped in with valuable insights and examples throughout. Selma Mehlsen Lau and Isak Mehlsen

Lau likewise pointed me to important examples of trading wits for vanity. Camilla Mehlsen, my wife and the love of my life, encouraged me to write the book. My Lady Day dropped the title on me after I had been commiserating about suckerhood repeating over and over again—testing her patience to the point of boredom—that ...

everything is a hustle.

References

—Abreu, P. & Brunnermeier, M. (2003). "Bubbles and Crashes", *Econometrica*, 71(1): 173–204.

—Akerlof, G.A. and Shiller, R.J. (2015). *Phishing for Phools: The Economics of Manipulation and Deception.* Princeton, NJ: Princeton University Press.

—Aral, S. (2020). *The Hype Machine: How Social Media Disrupts Our Elections, Our Economy, and Our Health — And How We Must Adapt.* Cambridge: MIT Press.

—Arash, A. Ouzan, S. & Shahzad, S.J.H. (2022). "Bubbles across Meme Stocks and Cryptocurrencies", *Finance Research Letters*, Volume 49, 2022: 103155.

—Ariely, D. (2008). *Predictably Irrational.* New York: HarperCollins.

—Aristotle (2009). *Nicomachean Ethics.* Trans. W.D. Ross. Oxford: Oxford University Press.

—Augustine (1991) *Confessions.* Trans. H. Chadwick. Oxford: Oxford University Press.

V. F. Hendricks, *Sucker Nation*,
https://doi.org/10.1007/978-3-032-22788-1

—Aquinas, T. (1981) *Summa Theologiae.* Trans. Fathers of the English Dominican Province. Westminster, MD: Christian Classics.

—Aumann, R.J. (1976). 'Agreeing to disagree', *Annals of Statistics*, 4(6), pp. 1236–1239.

—Aumann, R.J. (1999). *Interactive Epistemology I: Knowledge. International Journal of Game Theory*, 28: 263–300.

—Aumann, R.J. and Brandenburger, A. (1995) 'Epistemic conditions for Nash equilibrium', *Econometrica*, 63(5), pp. 1161–1180.

—Balaban, D.C., Szambolics, J. & Chirică, M. (2022). "Parasocial relations and social media influencers' persuasive power. Exploring the moderating role of product involvement", *Acta Psychologica*, Volume 230: 103731.

—Balcombe L (2025). "The Mental Health Impacts of Internet Scams", *Int. J. Environ. Res. Public Health*, 14; 22(6): 938–952.

—Banerjee, A.V. (1992) "A Simple Model of Herd Behavior", *Quarterly Journal of Economics*, 107(3): 797–817.

—Bernheim, B.D. (1984) .'Rationalizable strategic behavior', *Econometrica*, 52(4): 1007–1028.

—Berger, J. (2016). *Contagious: Why Things Catch On.* New York: Simon & Schuster.

—Berger, J. (2017). *Invisible Influence: The Hidden Forces that Shape Behavior*. New York: Simon & Schuster.

—Berger, J., Moe, W. W., & Schweidel, D. A. (2023). "What Holds Attention? Linguistic Drivers of Engagement", *Journal of Marketing*, 87(5): 793–809.

—Berger, L.M., Kerkhof, A. Mindl, F. & Münster, J. (2025). "Debunking "fake news" on social media: Immediate and short-term effects of fact-checking and media literacy interventions", *Journal of Public Economics*, 245: 05345.

—Bergmann, W. (1988). "Public beliefs about Anti-Jewish attitudes in West Germany: A case of 'pluralistic ignorance' ", *Patterns of Prejudice*, 22(3): 15–21.

—Betakova, D., Boomgaarden, H. & Lecheler, S. (2025). "The Role of Choice Architecture in Mitigating News Avoidance", *Digital Journalism*: 1–19.

—Bicchieri, C. (2017). *Norms in the Wild: How to Diagnose, Measure, and Change Social Norms*. Oxford: Oxford University Press.

—Bikhchandani, S., Hirshleifer, D. and Welch, I. (1992). "A Theory of Fads, Fashion, Custom, and Cultural Change as Informational Cascades", *Journal of Political Economy*, 100(5): 992–1026.

—Bjerring, J.C., Hansen, J.U., Pedersen, N.J.L.L. (2014). "On the rationality of pluralistic ignorance," *Synthese,*

191(11): 2445–2470.

—Bourdieu, P. (1988). *Homo Academicus.* Stanford: Stanford University Press.

—Brager, S. (2025). "As Social Platforms Withdraw Support, Professional Fact Checkers Fight Disinformation and 'Systemic Assault' on Facts", *Global Investigative Journalism Network*, April 1.

—Brandenburger, A. (1992). "Knowledge and Equilibrium in Games", *Journal of Economic Perspectives*, 6(4): 83–101.

—Broda, E., & Strömbäck, J. (2024). "Misinformation, disinformation, and fake news: lessons from an interdisciplinary, systematic literature review", *Annals of the International Communication Association*, 48(2): 139–166.

—Camus, A. (1942/2000). *The Myth of Sisyphus.* Translated by Justin O'Brien. Penguin Modern Classics. London, England: Penguin Classics.

—Cao, H. & Liu, S. (2022). "The Consideration of Meme Stock: Case Study of GameStop Saga", *Proceedings of the 2022: 7th International Conference on Financial Innovation and Economic Development (ICFIED 2022)*, 211: 2352–5428.

—Chin-Chiu Tan, T. & da Costa Werlang, S. R. (1988). "The Bayesian foundations of solution concepts of games",

Journal of Economic Theory, Volume 45, Issue 2: 370–391.

—Cialdini, R. (2007). *Influence: The Psychology of Persuasion,* HarperCollins Publishers.

—Cooperman, S.R. & Brandão, R.A. (2024). "AI assistance with scientific writing: Possibilities, pitfalls, and ethical considerations, *Foot & Ankle Surgery: Techniques, Reports & Cases*, Volume 4, Issue 1: 100350.

—Couldry, N. and Mejias, U.A. (2019). *The Costs of Connection: How Data Is Colonizing Human Life and Appropriating It for Capitalism.* Stanford, CA: Stanford University Press.

—Crypto.com (2026): "Exit Liquidity", *Crypto Glossary*: verified 20.01.2026: `https://crypto.com/eea/glossary/exit-liquidity`

—Curtis, A. & Fargher, N. L. (2013). "Does Short-Selling Amplify Price Declines or Align Stocks with Their Fundamental Values?", *Management Science*, 15.

—De Freitas, J., Thomas, K., DeScioli, P. & Pinker, S. (2019). "Common knowledge, coordination, and strategic mentalizing in human social life," *Proc. Natl. Acad. Sci. U.S.A.* 116 (28): 13751–13758.

—Douglas, K.M., Sutton, R.M. and Cichocka, A. (2017). "The psychology of conspiracy theories", *Current Directions in Psychological Science*, 26(6): 538–542.

—Douglas, M. (1990). *The Disciplined Trader: Developing Winning Attitudes.* Prentice-Hall.

—Dove, M. (2021). *The Psychology of Fraud, Persuation and Scam Techniques: Understanding What Makes Us Vulnerable.* Routledge.

—Espíndola, J. & Ramírez, J. (2025). "Kant's Spies", *Kant-Studien*, 116: 149–173.

—Fanelli, D. (2010). "Do pressures to publish increase scientists' bias?", *PLoS ONE*, 5(4), e10271.

—Farivar, S., Wang, F. Yuan, Y. (2020). "Opinion leadership vs. para-social relationship: Key factors in influencer marketing", *Journal of Retailing and Consumer Services*, Volume 59: 102371.

—Franck, G. (2018). "The economy of attention", *Journal of Sociology*, *55*(1): 8–19.

—Franck, G. (2020). *Vanity Fairs: Another View of the Economy of Attention.* Cham: Springer Nature.

—Frankfurt, H. G. (2005). *On Bullshit.* Princeton, NJ: Princeton University Press.

—Festinger, L. (1957). *A Theory of Cognitive Dissonance.* Stanford: Stanford University Press.

—Galeazzi, P., Pacuit, E. & Roy, O. (2026). *Epistemic Game Theory.* Cambridge Elements: Cambridge University Press.

—Gallegos-Erazo, F. A. (2024). "The Myth of Profitable Day Trading: What Separates the Winners from the Losers?", *SSRN*, August.

—Galloway, S. (2017). *The Four: The Hidden DNA of Amazon, Apple, Facebook and Google.* London: Random House Large Print.

—Gandini, A., Pais, I. (2020). "Reputation and Personal Branding in the Platform Economy" in Taylor, S., Luckman, S. (eds.) *Pathways into Creative Working Lives.* Palgrave Macmillan, Cham.

—Gennaioli, N., Shiller, R.J. and Shleifer, A. (2018) *A Crisis of Beliefs: Investor Psychology and Financial Fragility.* Princeton, NJ: Princeton University Press.

—Gigerenzer, G. (2008) *Rationality for Mortals: How People Cope with Uncertainty.* Oxford: Oxford University Press.

—Gigerenzer, G. (2018). "The Bias Bias in Bahavioral Economics", *Review of Behavioral Economics* 5: 303–336.

—Gogoi, G, Deimaphishisha, W. & Longkumer, S. T. (2024). "The Information Paradox in the Digital Environment", *SSRN*, June 20.

—Goodhart, C. (1975). "Problems of Monetary Management: The UK Experience", *Papers in Monetary Economics*, Volume 1, Reserve Bank of Australia.

—Guimaraes, B. & Pannella, P. (2024). "Short-covering bubbles", *Journal of Economic Theory*, Volume 219: 105846.

—Haimson, O.L., Tianxiao, L., Zhang, B.F. & Shanley, C. (2021). "The Online Authenticity Paradox: What Being "Authentic" on Social Media Means, and Barriers to Achieving It", *Proceedings of ACM on Human-Computer Interaction.* 5: 1–18.

—Halbesleben, J.R.B. & M.R. Buckley (2004). "Pluralistic Ignorance: Historical Development and Organizational Applications", *Journal of Management History* 42(1): 126–38.

—Harris, K.R. (2022). "Real Fakes: The Epistemology of Online Misinformation." *Philos. Technol.* **35**, 83–105.

—Harsanyi, J. C., & Selten, R. (1988). *A General Theory of Equilibrium in Games.* Cambridge: MIT Press.

—Hansen, P.G., Hendricks, V.F. & Rendsvig, R.K. (2013). "Infostorms", *Metaphilosophy*, vol. 44(3), April: 301–326.

—Harsin, J. 2015). "Regimes of Posttruth, Postpolitics, and Attention Economies," *Communication, Culture and Critique*, Volume 8, Issue 2, 1 June: 327–333.

—Haugen, F. (2021). The Facebook Files: A Wall Street Journal Invesigation, *The Wall Street Journal*, verified 05.01.2026.

—Hendricks, V.F. (2023). "All data is credit Information: Prediction, power and profit. That's the game-play of BigTech," *OECD Forum Network*.

—Hendricks, V.F. (2025). *What about me ism: Insights from Game Theory, Behavioral Economics and Moral Philosophy.* New York: Springer Nature.

—Hendricks, V.F. and Hansen, P. (2016) *Infostorms: Why Do We "Like"? Explaining Individual Behavior on the Social Net.* New York: Copernicus Books / Springer Nature.

—Hendricks, V.F. and Rendsvig, R.K. (2016). "The Philosophy of Distributed Information", in *Routledge Handbook of Philosophy of Information*, ed. Luciano Floridi, London: Routledge: 236–256.

—Hendricks, V.F. and Vestergaard, M. (2019) *Reality Lost: Markets of Attention, Misinformation and Manipulation.* Cham: Springer Nature.

—Hendricks, V.F, & Mehlsen, C. (2022). *The Ministry of Truth: Big Tech's Influence of Facts, Fictions and Feelings.* New York: Springer Nature.

—Hobbes, T. (1651/1996). *Leviathan.* Edited by R. Tuck. Cambridge: Cambridge University Press.

—Hofstadter, R. (1964). *The Paranoid Style in American Politics.* New York: Vintage.

—Hume, D. (1739/1978). *A Treatise of Human Nature.* Edited by L.A. Selby-Bigge and revised by P.H. Nidditch. Oxford: Oxford University Press.

—Hyland, D., Mittelmann, M., Murano, A., Perelli, & Wooldridge, M. (2024). "Incentive Designs for Rational Agents, in *KR '24: Proceedings of the 21st International Conference on Principles of Knowledge Representation and Reasoning,* article 44: 464–474.

—Jakob, S. (2026). "Why Meme Stocks Still Matter", *The Wall Street Journal*, 16.01.2026.

—Jahn, L., Rendsvig, R.K., Flammini, A., Menczer, P. & Hendricks, V.F. (2025). "A perspective on friction interventions to curb the spread of misinformation.", *Nature, npj Complexity* **2**, 31 (2025). `https://doi.org/10.1038/s44260-025-00051-1`.

—Kahneman, D. and Tversky, A. (1979) "Prospect Theory: An Analysis of Decision under Risk", *Econometrica,* 47(2): 263–291.

—Kahneman, D. and Tversky, A. (1984). "Choices, values, and frames", *American Psychologist*, 39(4): 341–350.

—Kahneman, D. (2011) *Thinking, Fast and Slow.* New York: Farrar, Straus and Giroux.

—Kant, I. (1996) *The Metaphysics of Morals.* Cambridge: Cambridge University Press.

—Kant, I. (1785/2012) *Groundwork of the Metaphysics of Morals.* Translated and edited by M. Gregor and J. Timmermann. Cambridge: Cambridge University Press.

—Kindleberger, C. P. & Aliber, R.Z. (2005). *Manias, Panics, and Crashes: A History of Financial Crises.* New York: John Wiley & Sons.

—Kipngetich, A. & Ochieng, P. (2025). "A Review of Online Scams and Financial Frauds in the Digital Age", *GSC Advanced Research and Reviews*, 22: 302–329.

—Keynes, J.M. (1936/2018) *The General Theory of Employment, Interest and Money.* London: Palgrave Macmillan.

—Krech, D. & Crutchfield, R.S. (1948). *Theories and Problems of Social Psychology.* New York: McGraw-Hill.

—Landwehr, C. & Schäfer, A. (2025). "The Promise of Representative Democracy: Deliberative Responsiveness", *Res Publica* **31**: 359–379.

—Lasch, C. (1979) *The Culture of Narcissism: American Life in an Age of Diminishing Expectations.* New York: W.W. Norton.

—Lefèvre, E. (1923). *Reminiscences of a Stock Operator.* Wiley.

—Lewis, D. (1969). *Convention: A Philosophical Study.* Cambridge: Harvard University Press.

—Lewis, D. (1996). "Elusive Knowledge", *Australasian Journal of Philosophy*, vol. 74(4): 549–567.

—Lewis, M. (2011). *The Big Short: Inside the Doomsday Machine.* New York: W. W. Norton & Company.

—Lewis, M. (2014). *Flash Boys.* New York: W. W. Norton & Company.

—Lynch, M. P. 1. (2016). *The Internet of us: knowing more and understanding less in the age of Big Data.* First edition. Liveright Publishing Corporation, a division of W.W. Norton & Company.

—Machiavelli, N. (1981). *The Prince.* Harmondsworth, UK; New York, N.Y. :Penguin Books.

—Mandeville, B. (1989). *The Fable of the Bees.* Indianapolis: Liberty Fund.

—Marker, S.L. & Hendricks, V.F. (2019). *Os og dem: Identitetspolitiske akser, ideer og afsporede debatter.* København: Gyldendal.

—McCarty, N., Poole, K.T. & Rosenthal, H. (2013). *Political Bubbles.* Princeton: Princeton University Press.

—Martin, B.A.S., Chrysochou, P., Strong, C.A. & Mills, A.J. (2025). "Cryptocurrency Research: A Conceptual Model for Future Research", *Psychology & Marketing*, 42: 2996-3004.

—Marx, K. (1867). *Capital: A Critique of Political Economy.* New York, NY: Cosimo.

—McNamee, R. (2020). *ZUCKED: Waking up to the Facebook Catastrophe.* New York: Penguin Books.

—Merton, R. K. (1968). "The Matthew Effect in Science", *Science*, 159: 56–63.

—Merton, R.K. (1973). *The Sociology of Science.* Chicago: University of Chicago Press.

—Muller, J.Z. (2018). *The Tyranny of Metrics.* Princeton: Princeton University Press.

—Murphy, J. (1998). *Technical Analysis of the Financial Markets. A Comprehensive Guide to Trading Methods and Applications.* Prentice Hall.

—Musk, E. in Rushkoff, D. (2023). "'We will coup whoever we want!': The unbearable hubris of Musk and the billionaire tech bros", *The Guardian*, 25.11.2023.

—Myerson, R. B. (1991). *Game Theory: Analysis of Conflict.* Harvard University Press.

—Nash, J. (1951). "Non-Cooperative Games", *Annals of Mathematics*, 54(2): 286–295.

—Nature (2010). "Publish or perish" *Nature,* Editorial **467**, 252: https://doi.org/10.1038/467252a

—Nietzsche, F. (1994) *On the Genealogy of Morals.* Trans. C. Diethe. Cambridge: Cambridge University Press.

—Nietzsche, F. (1844-1900/2006). *Thus Spoke Zarathustra: A Book for All and None.* Cambridge: Cambridge University Press, 2006.

—Ostrom, E. (1990). *Governing the commons: The evolution of institutions for collective action.* Cambridge, UK: Cambridge University Press.

—Paskalev, V. (2020), "Collective Reason, the Rationality Gap, and Political Leadership", *Ratio Juris*, 33: 169–195.

—Pearce, D. (1984). "Rationalizable Strategic Behavior and the Problem of Perfection", *Econometrica*, Vol. 52, No. 4: 1029–1050.

—Pinker, S. (2025). *When Everyone Knows that Everyone Knows: Common Knowledge and the Science of Harmony, Hypocrisy and Outrage.* London: Allen Lane.

—Plato (2007). *The Republic.* Translated by Desomond Lee, 2nd ed., Penguin.

—Plato (2015). *Theaetetus and Sophist.* Translated by C.J. Rowe, Cambridge: Cambridge University Press.

—Polyani, M. (1958). *Personal Knowledge: Towards a Post-Critical Philosophy.* Chicago: University of Chicago Press.

—Postman, N. (1985). *Amusing Ourselves to Death: Public Discourse in the Age of Showbusiness.* New York: Penguin Books.

—Prentice, D.A, Miller, D.T. (1993). "Pluralistic ignorance and alcohol use on campus: Some consequences of misperceiving the social norm", *Journal of Personality and Social Psychology*, 64(2): 243–256.

—Rousseau, J.-J. (1755/1997) *Discourse on the Origin and Foundations of Inequality Among Men.* Translated by D.A. Cress. Cambridge: Cambridge University Press.

—Sartre, J.-P. (2003). *Being and Nothingness.* Translated by Hazel Estella Barnes. 2nd ed. Routledge Classics. London, England: Routledge.

—Schelling, T. (1990). *The Strategy of Conflict.* Cambridge: Harvard University Press.

—Schneider, S. (2025). "Technocracy without society: A critique of nudging as an approach to managing risk," *International Review of Applied Economics*, 39(6), 919–941.

—Schüll, N.D. (2012). *Addiction by Design: Machine Gambling in Las Vegas.* Princeton: Princeton University Press.

—Shahrzadi, L., Mansouri, A., Mousa A. & Shabani A. (2024). "Causes, consequences, and strategies to deal with information overload: A scoping review", *International Journal of Information Management Data Insights*, Volume 4, Issue 2, 2024:100261.

—Shi, L. (2022). "Spillovers between Bitcoin and Meme stocks", *Finance Research Letters*, Volume 50: 103218.

—Shiller, R.J. (2000). *Irrational Exuberance.* Princeton, NJ: Princeton University Press.

—Shiller, R.J. (2017). 'Narrative economics', *American Economic Review*, 107(4): 967–1004.

—Simon, H.A. (1971). 'Designing organizations for an information-rich world', in Greenberger, M. (ed.) *Computers, Communications, and the Public Interest.* Baltimore, MD: Johns Hopkins University Press: 37–72.

—Simon, E. (2025). "What We Must Understand About the Dark Enlightenment Movement", *Time*, 24.05.2025.

—Sinderman, C.J. (2008). *Winning the Game Scientists Play.* New York: Basic Books.

—Sophocles (c. 441 BCE/2003). *Antigone.* Translated by R. Whittington. Cambridge: Cambridge University Press.

—Sophocles (c. 429 BCE/2008) *Oedipus the King.* Translated by S. Goldhill. Cambridge: Cambridge University Press.

—Srnicek, N. (2016) *Platform Capitalism.* Cambridge: Polity Press.

—Sunstein, C. (2009a). *Going to Extremes: How Like Minds Unite and Divide.* New York: Oxford University Press.

—Sunstein, C.R. (2009b). *On Rumors: How Falsehoods*

Spread, Why We Believe Them, What Can Be Done. New York: Farrar, Straus and Giroux.

—Sunstein, C.R. and Vermeule, A. (2009). "Conspiracy theories: Causes and cures", *Journal of Political Philosophy*, 17(2): 202–227.

—Taplin, J. (2017). *Move fast and break things: How Facebook, Google and Amazon Cornered Cultue and Undermind Democracy.* New York: Little, Brown and Company.

—Taylor, A., S. (2022). *Authenticity as Performativity on Social Media.* Palgrave Macmillan.

—Taylor, C. (1991) *The Ethics of Authenticity.* Cambridge, MA: Harvard University Press.

—Thiel, P. (2025). "Peter Thiel, the libertarian billionaire waging war on government", *Le Monde*, 22.07.2025.

—Thomas, K., DeScioli, P., Haque, O.S., & Pinker, S. (2014). "The Psychology of Coordination and Common Knowledge," *Journal of Personality and Social Psychology* 107 (4): 657–676.

—Thaler, R.H. and Sunstein, C.R. (2008) *Nudge: Improving Decisions about Health, Wealth, and Happiness.* New Haven: Yale University Press.

—Unkelbach C., Koch A., Silva R.R. & Garcia-Marques T. (2019). "Truth by Repetition: Explanations and Implications", *Current Directions in Psychological Science*, 28(3): 247–253.

—Urquhart, A., & Yarovaya, L. (2024). "Cryptocurrency research: future directions", *The European Journal of Finance*, 30(16): 1849–1854.

—van Lee, H.S. (2017). "A Formalization of the Greater Fools Theory with Dynamic Epistemic Logic", in Baltag, A., Seligman, J., Yamada, T. (eds) *Logic, Rationality, and Interaction*. LORI 2017. *Lecture Notes in Computer Science*, vol 10455. Berlin, Heidelberg: Springer.

—van Prooijen, J.-W. & Douglas, K.M. (2018). "Belief in conspiracy theories: Basic principles of an emerging research domain", *European Journal of Social Psychology*, 48(7): 897–908.

—Varoufakis, Y. (2024). *Technofeudalism: What Killed Capitalism*. New York: Vintage.

—Vestergaard, M. (2019). *Digital totalitarisme*. København: Informations Forlag.

—Vogel, H. (2009). *Financial Market Bubbles and Crashes*. Cambridge University Press.

—Vosoughi, S., Roy, D. & Aral, S. (2018). "The Spread of True and False News Online", *Science*, Vol. 359, Issue 6380: 1146–1151.

—Waterman B.R., Lubowitz J.H., Brand J.C. & Rossi M.J. (2023). "Strategies to Mitigate Against

Least-Publishable Units in Medical and Scientific Research Publications", *Arthroscopy*, November 39(11): 2255–2256.

—Webster, J. G. (2014). *The Marketplace of Attention.* Cambridge: The MIT Press.

—Williams, B. (1981) *Moral Luck.* Cambridge: Cambridge University Press.

—Wu, T. (2017). *The Attention Merchants: The Epic Scramble to get Inside our Heads.* New York: Knopf.

—Zhang, M. & Zheng, J. (2017). "A robust reference-dependent model for speculative bubbles", *Journal of Economic Behavior & Organization*, Volume 137: 232-258.

—Zuckerberg, M. (2019). "Zuckerberg Says Safeguarding Democracy is "Above our Pay Grade", *Vanity Fair*, 27.06. 2019.

—Zuboff, S. (2019). *The Age of Surveillance Capitalism.* London: Profile Books.

9783032227874